Surviving Orbit the DIY Way

Sandy Antunes

O'REILLY®

Beijing · Cambridge · Farnham · Köln · Sebastopol · Tokyo

Surviving Orbit the DIY Way

by Sandy Antunes

Published by O'Reilly Media, Inc., 1005 Gravenstein Highway North, Sebastopol, CA 95472.

O'Reilly books may be purchased for educational, business, or sales promotional use. Online editions are also available for most titles (*http://my.safaribooksonline.com*). For more information, contact our corporate/institutional sales department: 800-998-9938 or *corporate@oreilly.com*.

Editor: Brian Jepson
Production Editor: Melanie Yarbrough
Cover Designer: Karen Montgomery
Interior Designer: David Futato
Illustrator: Rebecca Demarest

August 2012:   First Edition

Revision History for the First Edition:

2012-08-23   First release

See *http://oreilly.com/catalog/errata.csp?isbn=9781449310622* for release details.

ISBN: 978-1-449-31062-2

[LSI]

Contents

Preface

Just how harsh is the space environment into which you thrust your DIY satellite? We look at what conditions your satellite must endure, how to test your satellite, and what launch, ground and orbit support you will need. In addition, we provide tips on making your overall plan and schedule, including the most important tests that will help your satellite survive and thrive in space.

Chapter 1 gives you a quick primer on what space and Low Earth Orbit (LEO) actually is, while Chapter 2 goes deeper with actual numbers that describe the conditions your satellite will be facing in space. Having set the scope of what space conditions are, Chapter 3 goes into more into detail about the radiation environment you'll be facing in the ionosphere and how to mitigate its risks. Chapter 4 provides you with a discussion of what testing is all about as well as offering some sample NASA CubeSat requirements and a smidgen of formal systems engineering.

In Chapter 5, we get into building a thermal vacuum chamber for mimicking the space environment. Chapter 6 walks you through building and running a vibration *shake* test to mimic the difficulties of your anticipated rocket launch. Chapter 7 guides you through high G-force testing of your payload using a homebuilt centrifuge.

Chapter 8 closes the book with a discussion on the usefulness of flight spares, guidelines on scheduling your tests, and help in choosing appropriate clean room or lab in which to test. In summary, the book tells you where your satellite is heading, what to test, and how and where to test it. We hope this book also serves you as a basic text on the Low Earth Orbit environment in general, so that you can design, build and test a spaceworthy picocraft.

By the end of this book, you will be ready to prove your picosatellite has the right stuff to deploy via a fiery rocket launch into the harsh vacuum of space. We also recommend the other books in this series: our primer on designing and building your craft in *DIY Satellite Platforms*, crafting your mission's science and technical goals in *DIY Instruments for Amateur Space*, and commanding, operating and downloading data from your satellite in *DIY Data Communications for Amateur Spacecraft*.

But first, let's really test the limits of the picosatellite you've built.

Conventions Used in This Book

The following typographical conventions are used in this book:

Italic

> Indicates new terms, URLs, email addresses, filenames, and file pass: [extensions].

`Constant width`

> Used for program listings, as well as within paragraphs to refer to program elements such as variable or function names, databases, data types, environment variables, statements, and keywords.

`Constant width bold`

> Shows commands or other text that should be typed literally by the user.

`Constant width italic`

> Shows text that should be replaced with user-supplied values or by values determined by context.

 This icon signifies a tip, suggestion, or general note.

 This icon indicates a warning or caution.

Using Code Examples

This book is here to help you get your job done. In general, you may use the code in this book in your programs and documentation. You do not need to contact us for permission unless you're reproducing a significant portion of the code. For example, writing a program that uses several chunks of code from this book does not require permission. Selling or distributing a CD-ROM of examples from O'Reilly books does require permission. Answering a question by citing this book and quoting example code does not require permission. Incorporating a significant amount of example code from this book into your product's documentation does require permission.

We appreciate, but do not require, attribution. An attribution usually includes the title, author, publisher, and ISBN. For example: "*Surviving Orbit the DIY Way* by Sandy Antunes (O'Reilly). Copyright 2012 Sandy Antunes, 978-1-4493-1062-2."

If you feel your use of code examples falls outside fair use or the permission given above, feel free to contact us at *permissions@oreilly.com*.

Safari® Books Online

Safari Books Online (*www.safaribooksonline.com*) is an on-demand digital library that delivers expert content in both book and video form from the world's leading authors in technology and business.

Technology professionals, software developers, web designers, and business and creative professionals use Safari Books Online as their primary resource for research, problem solving, learning, and certification training.

Safari Books Online offers a range of product mixes and pricing programs for organizations, government agencies, and individuals. Subscribers have access to thousands of books, training videos, and prepublication manuscripts in one fully searchable database from publishers like O'Reilly Media, Prentice Hall Professional, Addison-Wesley Professional, Microsoft Press, Sams, Que, Peachpit Press, Focal Press, Cisco Press, John Wiley & Sons, Syngress, Morgan Kaufmann, IBM Redbooks, Packt, Adobe Press, FT Press, Apress, Manning, New Riders, McGraw-Hill, Jones & Bartlett, Course Technology, and dozens more. For more information about Safari Books Online, please visit us online.

How to Contact Us

Please address comments and questions concerning this book to the publisher:

O'Reilly Media, Inc.
1005 Gravenstein Highway North
Sebastopol, CA 95472
800-998-9938 (in the United States or Canada)
707-829-0515 (international or local)
707-829-0104 (fax)

We have a web page for this book, where we list errata, examples, and any additional information. You can access this page at:

We have a web page for this book, where we list errata, examples, and any additional information. You can access this page at *http://oreil.ly/surviving_orbit_DIY*.

To comment or ask technical questions about this book, send email to *bookquestions@oreilly.com*.

For more information about our books, courses, conferences, and news, see our website at *http://www.oreilly.com*.

Find us on Facebook: *http://facebook.com/oreilly*

Follow us on Twitter: *http://twitter.com/oreillymedia*

Watch us on YouTube: *http://www.youtube.com/oreillymedia*

1/Life as a Satellite

...we've got to go out to Asteroid HS-5388 and turn it into Space Station E-M3. She has no atmosphere at all, and only about two per cent Earth-surface gravity. We've got to play human fly on her for at least six months, no girls to date, no television, no recreation that you don't de-vise yourselves, and hard work every day. You'll get space sick, and so homesick you can taste it, and agoraphobia. If you aren't careful you'll get ray-burnt. Your stomach will act up, and you'll wish to God you'd never enrolled. But if you behave yourself, and listen to the advice of the old spacemen, you'll come out of it strong and healthy, with a little credit stored up in the bank, and a lot of knowledge and experience that you wouldn't get in forty years on Earth.

— Robert Heinlein
"Misfit"

What's it like up there? Every 90 minutes, your satellite orbits the Earth. Each orbit passes high over a different geographic coordinate. The atmosphere you encounter is negligible, a residue of trace oxygen and other atoms with no real pressure to sustain you, just enough pressure to cause drag and (in months or years) reduce your orbit and cause reentry.

The Sun bathes you in heat and ultraviolet (UV) and X-rays and all the oth-er wavelengths of light. When in sunlight, your satellite heats up, perhaps uncontrollably. For half of each orbit, the Earth blocks the Sun and your satellite radiates out into that cold 3-degrees above absolute zero ambient temperature of space.

High energy particles (protons and electrons) stream from the Sun and interact with the Earth's magnetic field, creating beautiful aurora and po-tentially frying your electronics (see Figure 1-1). Very rarely, you might en-counter space dust or tiny bits of orbital debris.

Figure 1-1. *Space is beautiful, hostile, and survivable. Image of an aurora from the space shuttle, courtesy of NASA.*

All this assumes you survived the rocket launch.

Space is a harsh and unforgiving environment. It is harsh because it has no pressure or outside forces to provide structure, just vacuum. There is no air or liquid to conduct heat or buffer temperatures. It is filled with electromagnetic and particle radiation. Worse, if something breaks, you can't pop over to fix it. You get one shot and have to make that one shot work. Fortunately, through testing, you can practice your shots before the real event.

Space Is...

Space:

- is airless
- yet has atmospheric drag
- is hot and cold
- is insulating
- has electric and magnetic fields (as well as field *hot spots* like the South Atlantic Anomaly, where the Van Allen belts dip close to Earth)
- is influenced by solar weather
- lets the spacecraft jitter in attitude, bounce, and overall momentum
- requires orbital maneuvering rather than line-of-sight travel

Space is the highest of the three final frontiers (see "Frontiers" (page 3)). It is an airless, inhospitable place where your three biggest problems are a lack of air, extremes of hot and cold, and a surplus of electromagnetic energy. It also takes an enormous amount of fuel to get up there. To succeed in space, you must be lightweight, pack small, and be tough. This book tells you how to test whether your satellite will survive low earth orbit (LEO).

The air pressure in LEO is effectively nil. There's just enough air to provide atmospheric drag to de-orbit your satellite, but not nearly enough air to provide any structural or navigational assistance.

In space, there is no thermal blanket of air or water to help you retain or dissipate heat. Therefore, when the sun shines, you heat up—rapidly. When you are in darkness, you cool through radiative heat, and can keep cooling (in theory) until you reach 3K, or about -270 C or -454 F (for this work we'll be using Celsius).

To this environment, we add a fourth challenge—the rumbles required to get there. A rocket launch will apply intense g-forces, where acceleration of the rocket presses your payload as if it were feeling strong gravity. The rocket will vibrate and shake at different rates as it goes through different stages. Finally you get a *thump* as your satellite is ejected from the final stage. Here is where your satellite may drop a few inches very quickly. In any part of this, there may be mechanical mishaps as well.

Frontiers

The frontiers of space, the deep oceans, and the mind compete for the title of final frontier, but space got the official nod from Star Trek. "Space, the final frontier" is an extremely famous line from the opening of the original "Star Trek". The brain is often called "the final frontier of science", and ocean documentaries often dub the deep sea "our final frontier" or, in a bit of one-upmanship from the magazine COSMOS, "the real final frontier". In a humorous twist, death gets called "the undiscovered country" by Shakespeare, then re-appropriated by Star Trek as peace being the undiscovered country. In this context, we could call DIY Space the first, best hope for exploring one of the final frontiers. Hopefully peacefully, or at least with as little death as possible.

Launch Trouble

Looking at what can go wrong with a rocket launch, we're hard put to find something that can't go awry. Here's a short list of just some of the things that can go wrong during launch and early orbit checkout. You might have a launch vehicle problem and get into an improper orbit. Or, the forces of launch can damage your satellite from acoustics, g-forces, vibration, changes in atmospheric pressure, temperature changes, or outgassing. You may reach orbit but not have the right facing or spacecraft *attitude* (not an emotion, just means *facing and orientation*), making it difficult to communicate or misaligning your solar panels to the sun so you can't get power. Your satellite may underperform or overperform. You may miss detecting time-critical anomalies (satellite bits that go wrong) or situations (events that go wrong). Your team can make a hasty or incorrect decision. Your operating procedures may be incorrect. Your hardware or software may have bugs. (Thanks to Squibb, Boden and Larson, *Cost-Effective Space Mission Operations*, 2nd edition, pg 304, for this list.)

If all this sounds pessimistic, remember this is the book on testing. Your plan is to perform reasonable tests to mimic the conditions the satellite will face when you deploy for real. Good testing reduces risk, though you can never eliminate it. In fact, you need to be sure you don't get overzealous with your testing and actually break something for no other reason than testing it.

For example, if you want to test whether your car door is closing properly, you may want to slam it closed two or three times. But slamming it fifty times won't give you more information—and may break something that was working perfectly.

So our mantra is to devise a thorough test plan, carry it out in a reasonable fashion, analyze your test data, then stop. Each of those four steps is important.

How High Is Space?

How high is space, how far can you fall with a parachute, where do Low Earth Orbit (LEO) satellites reside, and where does the hard radiation from the sun get nasty? Gathered for the first time in one place is our High Altitude Explorer's Guide (Figure 1-2).

A typical airplane cruises at 9 km (6 miles) up, around 30,000 feet. Military jets (from the SR-71 onward to modern planes) can hit over 30 km (19 miles) up, over 100,000 feet. You can parachute from that height. In 1960, Joseph Kittinger set the record at 31.3 km (19.5 miles), or 102,800 feet. Felix Baumgartner is planning to use a rocket to freefall from 36 km (over 22 miles)—an 118,000 feet fall—some time before 2015.

But those aren't *space* (Chapter 2). In the US, "space" begins at 80.4 km (50 miles), or 264,000 feet. General international consensus sets a similar limit for the start of space as 100 km (62 miles), or 380,000 feet.

"Low Earth Orbit" (LEO), where many satellites live, goes from 160 km (100 miles, 525,000 feet) to 2,000 km (1,240 miles, 6.5 million feet). My own Project Calliope picosatellite will be 230 km up (143 miles, 755,000 feet). The International Space Station (ISS) cruises higher up, from 278 km (173 miles, 912,000 feet) to 460 km (286 miles, 1.5 million feet).

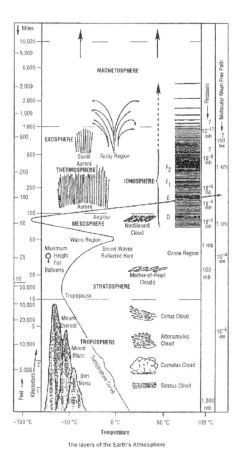

The layers of the Earth's Atmosphere

Figure 1-2. *Layers of the Earth's atmosphere (courtesy NASA/MSFC)*

Starting above the space limit but a bit before LEO, the inner Van Allen Belts, which magnetically shield the Earth's surface from high energy particles, extend from 100 km (62 miles, 33,000 feet) up to 10,000 km (6,200 miles, 3.3 million feet).

Finally, geostationary orbits are at 35,786 km (22,236 miles, 117.5 million feet). These geosynchronous orbits, lined up above the Earth's equator, have an orbital period equal to one day, so they hover over the same spot of the Earth.

Together, that's "space", with an emphasis in this book on LEO. We'll walk through several bits of homemade gear to tackle these tests. For a vacuum test, you can make a simple chamber using a pressure cooker and an automotive brake vacuum pump. For thermal vacuum, you need to add an IR heat source. For vibration and shake testing, you can modify power tools to do simple test scenarios. G-forces can be tested primitively using the centrifugal force provided by you swinging a rope, or with more technical elan using a power drill and a long pair of swing arms. As for the drop test, that's both the simplest and most nerve-wracking, as you simply have to be willing to drop your work. But first, a deeper look at space. Let's look at the hard numbers for where we'll be sending our picosatellite.

2/The Measure of Space

By the numbers, we're 1.496 x 10^8 km from the Sun, aka 1 Astronomical Unit (AU). It takes about 8 minutes for light or any kind of radiation from the Sun to reach the Earth. Some events—like plasma thrown off from the Sun—can take 1-4 days to reach us. The sun throws out light (radio through gamma) as well as energetic electrons and protons, and big wind-like streams of hot hydrogen gas (solar plasma), as shown in Figure 2-1.

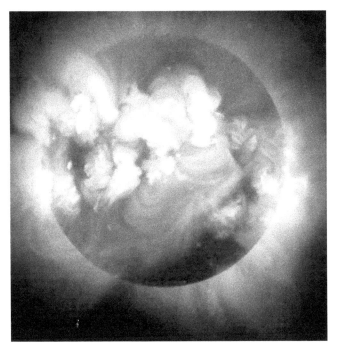

Figure 2-1. *Yokkoh image of the Sun in UV showing its active nature (courtesy JAXA)*

While Earth's atmosphere is mostly nitrogen, up at Low Earth Orbit (LEO) has almost no atmosphere at all. We get the worst of both really—no air to use for breathing, but just enough molecules of air to cumulatively cause drag and eventually make our LEO satellites de-orbit.

Our atmosphere blocks most of the harder radiation the Sun puts out—X-rays and UV (especially, for UV, the Ozone Layer). By definition what we see on Earth is *visible light*, plus a little light UV and a fair amount of invisible radio emission. All that blocked stuff is visible from orbit, however.

In fact, that's why we launch satellite telescopes—less atmosphere to block any kind of light.

Note some plastics (polymers) are sensitive to UV—they break down in the presence of UV. Fine on Earth, but terrible for space-going hardware. ESA maintains a large list of materials and their responses to the space environment in their ESA materials database at *http://esmat.esa.int/*. In general, if it's a material that has a strong smell or that oxidizes (rusts) easily, it's a poor choice for space.

At sea level, our atmosphere has a pressure of 1×10^5 Pascals (Pa), with a density of 1.225 kg/m^3 (and 10^{24} molecules/m^3, if you like counting molecules), assuming a moderate reference temperature temperature of 288 K (15 C). The *US Standard Atmosphere* model (Figure 2-2) illustrates the temperature, density and pressure variation as you increase in altitude.

By 86 km up (which is where the US Standard Atmosphere model stops calculating), you are facing (at 187 K/14 C) pressure of 0.3 Pa (3×10^{-5} atmospheres) and only 0.000006 kg/m^3 of air mass. At that height, you're still not officially in *space* and would not count as an astronaut if you flew there. Space is typically defined as 100 km up and higher. Putting these together in comparison, 99.9% of the Earth's atmosphere is in the layers below 50 km, while only 0.1% of the atmosphere in the 50 km to 400 km range.

For LEO, in the range of 300-900 km, the atmosphere is thin (low density) but not zero. LEO is in the part of the atmosphere called the Thermosphere, and the Thermosphere contains the region called the Ionosphere. The Ionosphere is where auroras occur. By 400 km up, you have about 10^{14} molecules per cubic meter (10^{-10} kg/m^3), with an equivalent pressure of, for all practical purposes, nil (10^{-5} Pa and lower).

Pressure is hard to define in the ionosphere, since pressure is a measure of air molecules impacting a given object, and there just aren't many molecules at that height. Put another way, any individual molecule will typically travel 1 km or so before colliding with another molecule. The ionosphere's density also varies with solar activity and whether you are in sunlight or dark. As you will often read in this book, LEO is a vacuum in terms of harsh unsurvivability, with just enough gas molecules to cause oxidation and drag in the

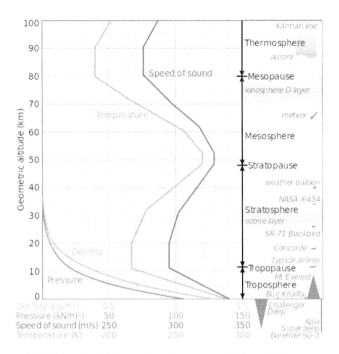

Figure 2-2. *US Standard Atmosphere (courtesy Wikimedia)*

long term. It's the worst of both worlds: no air to use, just enough to eventually force reentry. Most of that is oxygen, which means you have oxidation (corrosion) to worry about. Atomic oxygen is the major component of the residual LEO atmosphere (Figure 2-3).

By the time you reach geosynchronous height, 36,000 km up, the density is 10^{-20} kg/m^3 and the pressure is 10^{-15} Pa, and it pretty much stays at that low density as you travel within the solar system.

You heat up due to absorbing solar radiation (1371+- 5 W/m^2 or so, ref) as well as sunlight reflected off the Earth (Earth's albedo) plus Earth black-body emission, totaling perhaps 200 W/m^2. While the solar wind is at 2×10^5 K, it is so thin it provides no real heating.

The ionosphere by definition is where ionization of plasma occurs, and is the region above 86 km. The incoming solar radiation (called *incident solar UV*, a fancy term for the sun's ultraviolet light) disassociates the atmosphere into its individual elements, then disassociates the electrons from the nuclei. Its peak density is in the 300-400 km range, though the density of the plasma is still less than the density of the neutral molecules; both co-exist.

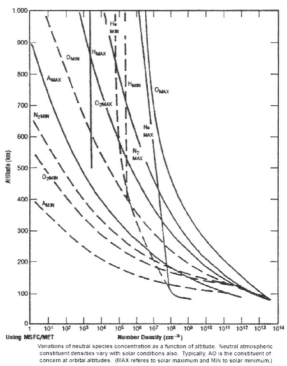

Figure 2-3. *What elements exist at different altitudes (courtesy NASA)*

Outgassing

Back to the lack of air. Every get something wrapped in plastic and it smells all plastic-y? You're sniffing some plastic molecules that outgassed from the wrapping material. Outgassing (also called sublimation) is the vaporization of a solid going to a gas state due to low pressure. At the typical 10-11 to 10-15 Pa of LEO, many materials stable in an Earth lab will outgas. Heating increases outgassing.

Outgassing has two problems; it erodes the material that is outgassing, true —but more crucial, the outgassed material may condense on—and thus coat —other surfaces. This can change their conductance or, for optics and detectors, coat them so they no longer can see.

For this reason, do not put lubricants into space. Ground-based lubricants have a high vapor pressure and outgas quickly. They therefore are terrible as lubricants, since they evaporate off your moving parts, and terrible as a contaminant, since they'll coat random surfaces.

Orbital Thermal Profiles

In space the temperature can range from -269 C (4K absolute background) to over 400 C. Low Earth Orbit has some cooling/heating cycles as you are in and out of sunlight, and you can expect a floating object to experience a range from -160 C to 200 C (as a rough estimate). By clever thermal design —described below, and using tricks like rotating your craft to distribute heat more evenly and using insulation, you can cut this to a range of -100 C to 100 C. This is still outside the range most off-the-shelf electronics can handle.

If you minimize your radiative properties (choosing white or reflecting casings) to reduce both heating due to the sun and re-radiating that heat in the dark, and you assume there is some on-board heating of your components due to being in use, you can achieve a temperature range of -30 to 90 degrees as typical. Why are these typical? Those are often the limits that off-the-shelf electronics function at, so we're going to assume you design your craft to keep to that temperature range.

For a more thorough answer on temperature profiles, I quote "IRStuff" from eng-tips.com (*http://www.eng-tips.com/viewthread.cfm?qid=102652*):

> An unpowered, unilluminated object, in deep space, will equilibrate to approximately 4 Kelvins. You can passively verify that temperature the same way that the COBE satellite measures the space background, e.g., through the blackbody emissions of the object in the microwave regime. In any region close to the Earth, the temperature can vary from about -160C to over 200C.

Complete thermal calculations are very complex (see sidebar "Environment and Thermal References" (page 11)) and, for a picosatellite, perhaps overkill. One advantage of small size is that you can assume the satellite is approximately a black box that equally absorbs and radiates uniformly.

Magnetic Fields

The Earth has a magnetic field; it's what makes compass needles point to-wards the poles. This field is also what protects us from the worst of the charged particles that stream from the Sun. The earth's magnetic field at the surface is due to the Earth's core, but in the ionosphere we have both the core field and reaction of the Earth's magnetosphere to the solar wind.

The magnetic field as a flux density is about 1T at the surface and drops only about a factor of two by 400 km. You have to get more than 2000 km up before you lose even half the magnetic field flux. For the full picture, there are very complex models (NASA models AE8 and AP8) that can predict the electron and proton (charged particle) fluxes for any LEO orbit during different parts of the 11-year solar cycle. Most of these charged particles are kept from LEO by the magnetosphere, but again, in LEO you will experience some of the particle fluxes.

Aurora occur at the poles, where the magnetic field lines dip (Figure 2-4). There is also the South Atlantic Anomaly, a region where the Earth's magnetic field lines dip lower and thus increase the electron/proton flux experienced in orbit. Charged particles can cause temporary and permanent damage.

Charged particles cause single event upsets (SEUs) that make random signals appear in your electronics. Charged particles also degrade solar arrays and electronic parts, and can dump charge into electronics to cause surges (dielectric charging). LEO is more dominated by protons, with electrons more common in higher orbits, but again this fluctuates and depends on solar activity. Protons do more damage, particle for particle, than electrons, as they are heavier and thus carry more momentum.

Orbital Mechanics

How objects move in orbit is key to understanding your satellite's environment and behavior. Put simply, for a given orbit, everything moves at the same speed regardless of mass. If you have the ISS, VelcroSat, and a cloud of debris in the exact same orbit, they'd all move at the same speed but never intercept each other.

Also, if you speed up in an orbit, you also change your orbital path to a higher one (because your centripetal force outwards is greater so you're overcoming more of gravity's downward pull). If you slow down in an orbit, you are giving in to gravity and your orbit decays.

To make it weirder, the speed you need to have in an orbit is lower the higher up you are—because again, you are fighting less gravity higher up, so you need a lower speed.

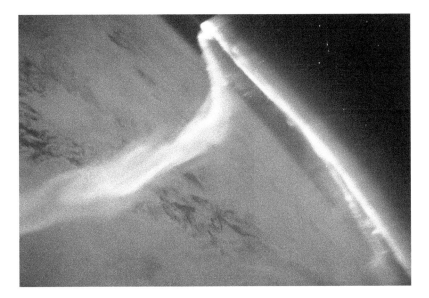

Figure 2-4. *An aurora viewed from orbit, and just the sort of thing you don't want an untested satellite to fly through (image courtesy NASA)*

How can this be—first I say "if you slow down, you go down" then I say "if you are higher up, your orbital velocity is slower". These seem contradictory, but they actually are two different problems.

Each orbit is like a rigid track—if you have a specific orbital radius, you have a fixed velocity for that orbit. And the higher the track, the slower the velocity you need to stay there. The track is a perfect balance of centripetal force and gravity.

However, to *transfer* to a different orbit, you have to jump up (or drop down)-- briefly. You have to either give yourself a kick to jump higher, or let gravity briefly take over. That extra kick up (extra energy) gets used up in moving you higher. Similarly, that loss of velocity (and thus loss of kinetic energy) lets gravity take over and drop you down.

So you spend energy to move up to a higher orbit but, once you get there, the total kinetic energy you need to stay there is lower than before.

Meanwhile, the Earth is rotating below, so the ground track or ground trace of the path of your orbit shifts relative to land (Figure 2-5). Over one day, an LEO satellite may pass over the Earth 15 times (Figure 2-6), each time with position shifting a bit. This also means that a fixed ground communications antenna will only be able to see a subset of the entire day's orbits.

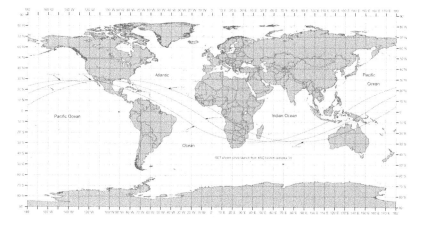

Earth parking orbit ground track for 3 orbits at a 90 degree launch azimuth from KSC

Figure 2-5. *Three successive orbits plotted for an LEO satellite. Each curved line is one orbit. Note how each orbit is over a slightly different set of geographic positions. Image courtesy NASA.*

Figure 2-6. *A full day's worth of orbit tracks for the now-gone UARS satellite, showing just how much ground an LEO satellite can cover in one day.*

In an elliptical (not circular) orbit, you move faster when you're closer to the Earth, and slower when you're further out. It's a stable orbit, but the speed varies. Also, how it's aligned to the Earth varies.

Orbital Debris

And now space debris. There's junk in space, and it can present a hazard for spacecraft (Figure 2-7). Are impacts an issue? Micrometeors and debris are infrequent and, for picosatellite work, negligible. If you're tiny and don't last long in orbit, they are the least of your problems. While a picosatellites might encounter space debris, our small size, hardiness, and short lifetime make it a possibility not worth accounting for.

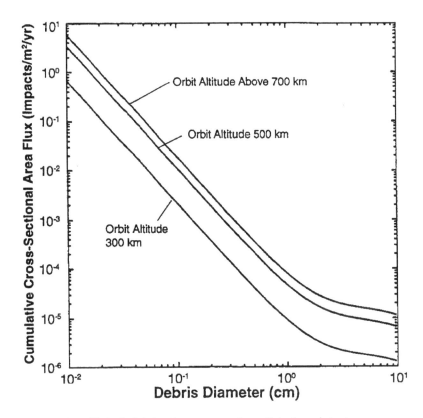

Figure 2-7. *Plot of debris size versus where it is found, just so you can worry.*

If you are expecting trouble, you can look up multilayered insulation (MLI) shields, such as used on the ISS. A common meteor shield design is the Whipple bumper, which has a thin shield layer that shatters the incoming particle, an empty gap, then a thicker inner shield layer that acts as a back-up wall. The first layer spreads out the incident particle so the second layer can stop it.

Capturing orbital debris is a problem DARPA and others are trying to solve, including using a "catcher's mitt" to clean things up. What this means in terms of maneuvering is we cannot just fire a rocket to fly faster and hope to catch debris that's ahead—because that would also change our orbit. It also means, in general, we can't say we'll fly *faster* than the stuff in our orbit and sweep stuff up. Instead, any change in speed also changes our orbit's altitude and/or shape (eccentricity).

Figure 2-8. *Simulated image showing debris in LEO up to 2000 km up (courtesy JSC/NASA).*

Since we can't just be like a pool net, sweeping out an orbit, how do we intercept debris? What makes this solvable is primarily that orbits aren't circular—especially for debris. The eccentric (squished circle) orbits therefore intersect in different ways, creating the debris problem.

However, picosatellites are an excellent testbed for trying out orbital debris concepts. This is why I'm happy to be advising the Capitol College's VelcroSat orbital debris removal prototyping picosatellite effort. Ergo, a debris-catcher can, via timing, work to intercept debris that is in a different orbit that has a brief overlap with the VelcroSat orbit. This is also why debris collisions occur—items in orbits of different eccentricity and orientation collide (Figure 2-8).

Now that we know what we are going to be facing, let's look at how we're going to schedule out our tests. After that, we'll get into building the test gear.

3/Space Radiation Environment

Picosatellites, like any Low Earth Orbit satellite (LEO), are going up to, well, LEO. Space weather—radiation and energetic particles emitted from an active Sun—can damage satellites. This region of space is partially protected from the worst effects of space weather by the Earth's ionosphere, but it is an active and threatening place.

If space wasn't active, there wouldn't be any point in sending up my own Project Calliope to measure it. However, we'd prefer to keep the physical damage to the electronics to a minimum. The primary source of damage due to solar activity is due to highly energetic electrons, protons, and ions emitted by the Sun (see Figure 3-1).

Figure 3-1. *Aurora caused by solar activity charging the Earth's ionosphere, as viewed from space. Image courtesy of NASA.*

The particle and radiation environment is not static. Besides time variation due to solar activity, the shape of the Earth's magnetic field also can lead to dips in the field that cause a higher hazard in certain geographic areas. The most known dips are where the field lines converge at the Earth's north and south poles. These are responsible for the aurora. There is also a dip near Brazil, called the South Atlantic Anomaly (SAA) (Figure 3-2) that LEO satellites can pass through during one or more orbits each day. Satellites with sensitive detectors often decide to either shut down or cease collecting data during the SAA passage, as the increased background can make the data too noisy to use (as well as risking damage to the satellite electronics).

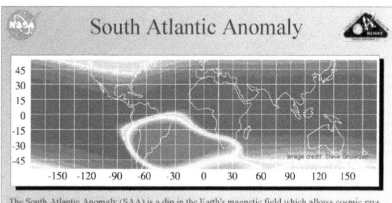

The South Atlantic Anomaly (SAA) is a dip in the Earth's magnetic field which allows cosmic rays and charged particles to reach lower into the atmosphere and interfere with communication with satellites, aircraft, and the Space Shuttle. The geologic origin is not yet known.

The enhanced particle flux in the SAA also strongly affects X-ray detectors, which are in essence particle detectors. The ROSAT PSPC had to be turned off during passage through the SAA to prevent severe damage. While the ROSAT HRI could be left on during the passage, it could collect no useful data. The light blue and green bands at the top and bottom of the image are due to an enhanced particle flux above Earth's auroral zones (particle belts).

Figure 3-2. *The South Atlantic Anomaly (SAA), a radiation/particle risk area for LEO satellites. Graphic courtesy of NASA.*

Space Weather Events

Space weather can kill astronauts. This is one of the motivations for funding space weather efforts. Solar events—flares, particle storms, and coronal mass ejections—can knock out GPS and cell phone reception, screw up radio and radar, and endanger airline pilots flying the polar routes. All of these damaging effects are well worth mitigating. But what about circumstances higher up?

In an article titled "Fake Astronaut Gets Hit by Artificial Solar Flare", NASA reports on their upcoming experiment to see just how much damage a solar flare would cause to an unprotected astronaut.

On Earth, we have the luxury of being able to do something, if given advance notice. Travelers can be advised about GPS and cell phone outages.

The military can plan around disruption. And airplanes can and do reroute when there's a chance of solar trouble. But what befalls an astronaut, up past the shielding we get from Earth's magnetic field, when they get notice of an event hours or minutes away? Where can they go?

I was told a possibly apocryphal take from the Apollo, on the instructions to astronauts if there should be a solar event while they were on the Moon. "Maximize shielding. Climb into the lowest point of the lunar lander. Have one astronaut lay on top of the other. The one on the bottom might live."

In 1972, there was a massive solar flare, between the Apollo 16 and 17 missions. NASA decided to see what would have happened if, oh, there had been astronauts up in an Apollo 16.5 at the time. Now, when I did work predicting how long CCD (charge-coupled device) detectors could last in space, I used Monte Carlo computer simulations, and there's only so far a simulation can take you when dealing with a complex system. And few things are more complex than the human body.

NASA's method for testing human survivability in high radiation environments is brutally clever. They created a dummy, filled it with human blood cells, then stuck it in front of the Brookhaven particle accelerator and simulated a solar proton storm. They took it out and measured the cell damage, refilled it, and repeated this test.

These particles can penetrate past the satellite's skin and the surface of the electronics and dump their energetic charge into the electronics itself. This can cause glitches—Single Event Upsets, where the electronics briefly get a wrong signal value. It can also degrade or erode the solar panels and other sensitive bits—though this is less of a factor for our short-lifetime (6 weeks, nominally) mission.

If a picosatellite was a person, the radiation could damage DNA and similar. For now, though, we can just shield our satellite from particles and be safe, correct?

Shielding

In a French word, *non*. Shielding can help protect electronics, but due to the actual mechanism that causes damage, shielding can also increase the risk of damage. Unlike armoring a medieval knight, where you want to add as much armor as they can carry, for satellites you need to understand how the potentially damaging particles interact with the spacecraft.

First, it is only specific particles, of a specific energy, that will react with the electronics. Particles of low energy will be shielding by the body of the spacecraft or even the paint and thin silicon layer coating the electronics. Like these incredibly detailed schematic show:

hmF2_b: height of 90% of the maximum electron density
 below hmF2

hmF2 : height of the maximum electron density

hmF2_a: height of 90% of the maximum electron density
 above hmF2

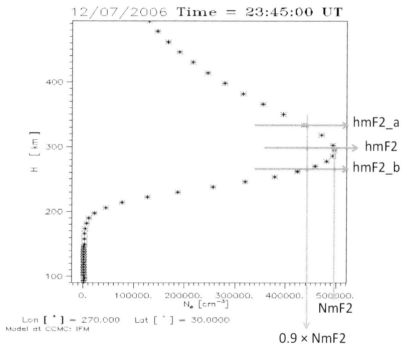

Figure 3-3. *Electron distribution versus altitude. If you really want to explore shielding, I figured this graph wouldn't scare you off. Figure courtesy of NASA.*

Particles (electron or proton or ion):

the electronic component:

their interaction (particle stopped):

Particles of very high energy will barely be slowed by the small amount of material and will typically pass straight through the satellite and electronics, reacting negligibly:

The damaging particles are, in a Goldilocks fashion, the particles at a *just right* energy. They have enough energy to pass through the blocking materials to get into the electronics, but not so much energy as to zip out. Instead, they deposit all their *impact* into the electronics, causing the SEU or damage:

To protect from these, first, you have to know what the range (or spectrum) of particles are in space. Here (Figure 3-4) are two sample spectra, of very different shapes. The energy of the particles range from some low *E0* up to higher values, ending at *E3*. One spectrum has more low energy particles than high energy, the other is the opposite (more high energy particles than low energy ones).

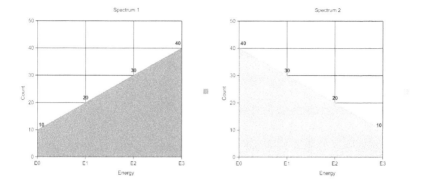

Figure 3-4. *Two sample incident damaging particle spectra. One spectra is mostly high-energy particles, the other mostly low-energy particles. We assume low energy particles get absorbed by your electronics, causing damage, but higher energy particles zip through with little effect. For Spectrum 1, adding shielding will end up attenuating the bulk of the particles and result in an increase in damage.*

Let's add a detail—say all particles of energy E0 or lower are blocked by the satellite as it is, and do no damage. And all particles of energy E2 or higher go straight through and cause little or no damage. The only particles that will damage our satellite are those of energy *E1*.

If the particles appear as per Spectrum 1, then we get around 20 of the damaging E1 particles. If Spectrum 2 is the case, we get around 30 of the damaging E1 particles. The damage the satellite receives obviously depends on which spectrum the Sun is emitting.

Now let's add shielding. Shielding attenuates, or slows down, all particles. If particles are of low energy, they get slowed down so much they are completely blocked. If particles have high energy, they lose some energy passing through the shielding, but still continue on—at a lower energy.

So now we add a block of shielding. Let's say it's enough to slow down each particle by 1 of our labeled energy levels. An E3 particle gets slowed down to become an E2 particle. E2s turn into E1s. E1s turn into E0s. E0s get blocked completely by the shielding.

What then happens to our satellite if we shield it? Suddenly, with shielding, everything at E1 or lower is blocked but the E2 particles get *downshifted* to E1 and cause damage (and the E3 get slowed to E2 and therefore still pass through without causing any trouble).

Under this, Spectrum 1 has 30 formerly E2 particles that get slowed by the shielding, *become* E1, and thus damage the electronics. With Spectrum 2, there are only 20 formerly E2 particles that turn into damaging E1 particles.

As a little table, then, we find that, without shielding, the satellite takes more damage if the Sun emits Spectrum 2, but with shielding, Spectrum 1 will do more damage. The choice of whether to add shielding or not therefore depends not just on the amount of shielding, but on understanding what the specific energies and amounts of particles the Sun is likely to emit. Adding more shielding can, in some cases, actually increase the damage and risk to your electronics.

For a larger space mission, you can obtain the predicted space environment and also create a 3D model of the *attenuation curves* for the materials and thicknesses of your satellite in order to create an estimate on particle damage. To be very accurate, you should run multiple Monte Carlo simulations of particles hitting the satellite, since real particles have a *probability of interaction* rather than the simple hit/miss model I give here. That is a subject for a future... no, wait, that was my Master's thesis way too long ago (later published as Antunes et al, 1992, in Experimental Astronomy, vol. 4, no. 2.)

History of Damage

Most picosatellites are not just limited in weight, but also don't need to really worry about short-term transient damage like SEUs. Instead, we'll just take our lumps and hope that the bulk of the data we download—itself just a fraction of the entire data captured—will suffice.

A case study of cumulative damage to imaging CCD detectors (back in 1992, by Antunes et al) discusses how the noise of the detectors increase as they suffer radiation damage. Electrons affect CCDs by increasing background rates and by producing ionization damage. Low energy electrons produce signal change indistinguishable from x-ray events. Ionization within the oxide insulating layers can produce flatband voltage shifts and increased contributions from surface state dark currents. Within one to two years, they lose 10^{-5} in charge transfer efficiency, which means (multiplied by the number of pixels you read per row, e.g., a 512 x 512 CCD would suffer 512×10^5 in additional noise). This substantiates our claim that, in general, a short-lived picosatellite does not have to significantly worry about the radiation environment for their typically shorter lifetimes.

Electronics Noise

In addition to the radiation environment provided by space, your satellite electronics are going to bring along their own noise and interference. One part of assembly is making sure all my circuits are properly shielded and not sending out interfering signals. A decent magnetometer—a meter to measure magnetic fields—costs $200-$700 dollars. While at the vitamin shop, I beheld a CellSensor, a device that measures and traces cell phone and power line RF (radio frequency) emission. It has a range of milliWatts (for RF radiation) and milliGauss (for magnetic fields). And it was discounted to $20 (Figure 3-5).

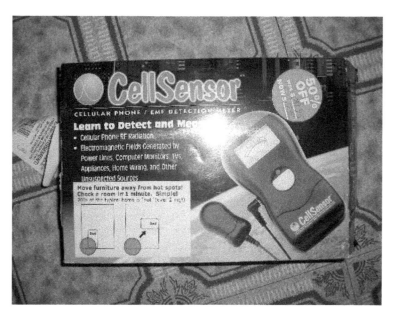

Figure 3-5. *A (commercial off-the-shelf (COTS) electromagnetic (EM) detector, sold at a health food shop to paranoid grounddwellers but repurposed for space work.*

So, for $20, I now have a new addition to my test rig and construction setup! One part of 'do it yourself'—in any domain—is reappropriating hardware for your own purposes. One person's fear of cell phones led to me saving several hundred $$ for my lab.

Meanwhile, up in space, I'm trying to estimate the magnetic field of the ionosphere. My current detector measures scales in 10s of Gauss, the ionosphere has a field strength on order of 0.3–0.6 Gauss, with fluctuations of 5% (plus we'll have orbital variations). Since what my Calliope picosatellite is measuring is the fluctuation, not the field strength, this means I need to capture 0.06–0.1 Gauss signals if I want to simulate a magnetic detector on Earth.

However, that's just based on uniform signal...we'll be in a polar orbit, and the field lines dip more there so the magnetic field (and variability) will be higher. Clearly I need to do some orbital simulations. I suspect I will need to either rewire or rework the existing sensors. It's a good thing I have a test rig —every $10 spent on the ground saves thousands in potential failed mission risks!

4/Testing Formalism

There is functional testing, as opposed to environmental testing, which is different from launch survivability testing—oh my, so many tests. Functional testing is where you verify whether a component works or not, either isolated or when integrated into your overall satellite. All the steps laid out previously provide you with a functional test schedule.

Environmental testing is where you test that a working component continues to work even when subject to the rigors of space. This primarily means vacuum testing and thermal testing. A homemade thermal vacuum chamber will do nicely. This testing should be done for each of the isolated components. Vacuum testing is a pain, because you had to:

1. Insert the part into your thermal vacuum chamber
2. Set up some test measurement system
3. Get to vacuum, run the test
4. Repressurize
5. Test the component once it comes out of thermal vacuum chamber, to ensure repressurizing didn't damage everything

I recommend you do thermal vacuum chamber testing after you've finished your functional testing. You will have to break down your full assembly and retest each component in isolation, which is tedious. However, you will already have verified your original design and assembly and integration, so the bulk of this book is the test of survivability. Further, because you know integration works, you can reduce the number of test steps while maintaining accuracy.

NASA CubeSat Requirements

NASA has one formal specification (among many) from their Launch Services Program titled "Program Level Poly-Picosatellite Orbital Deployer (PPOD) and CubeSat Requirements Document" (*http:// www.nasa.gov/pdf/627972main_LSP-REQ-317_01A.pdf*). This short document (nearly shorter than its title, at just 14 pages) gives a set of NASA specifications on the requirements for CubeSat and PPOD (CubeSat deployment racks). They include vibration, shock, and thermal requirements as well as broader categories like power limits during phases of launch. If your satellite matches these requirements, you can likely be permitted on a NASA rocket. If you fulfill the NASA requirements, you likely can fulfill any private launcher's requirements as well.

Among their requirements are several policy-level issues that are worth examining when you are designing your payload. The following are forbidden: pressurized vessels, propulsion systems, radioactive materials, explosive devices (including release bolts to open or expand mechanical assemblages). Several picosatellite teams are specifically using picosatellites to develop ion drive or pulsed plasma propulsion systems, so be aware that you may need to negotiate with your rocket provider if you are moving into territory this one set of NASA requirements forbids. That doesn't mean you can't do your experiment, but it does tell you that you will have to do additional justification and testing.

Other requirements are operational requirements. CubeSats under this spec must be powered off from the time of delivery until after deployed in-orbit (no power on during launch). Likewise, they cannot radiate RF (radiofrequency emission, either as radio signal or electronics noise) until 45 minutes after being deployed in-orbit. NASA under this spec presumes the payload is a CubeSat being deployed via a PPOD deployment mechanism. Again, you can do another plan, but the more you drift from this spec, the more paperwork you will incur.

In favor of picosatellites and standardization is that you can escape some of the risks that larger, more complicated missions face. Picosatellites are compact and minimalist payloads. I think most TubeSats and CubeSat are more resilant than a larger traditional satellite payloads (that might have extendable solar panels, optical benches, alignment requirements of 1mm, etc.), so such homebuilt tests in this book (I am hoping) will be sufficient pretests for most DIY satellite projects. Folks doing plasma drives or that require alignment can chime in with a rebuttal, since they're doing more mechanically precise work.

The primary limitation in home test gear is we cover only a partial span of the entire range of potential launch problems that can occur. For example, our homemade *shake table*, built with an orbital sander on a router control, can span vibration frequencies up to 250 Hz. While that covers only the lower end of all possible rocket launch vibrations, it does a good enough job at breaking poor soldering joints such that a payload that survives that will likely survive launch. Similarly, you can get 3G of false gravity from the home centrifuge we offer here—lower than the 10G limit a rocket potentially might have, but certainly good enough to do your primary structural testing.

Just as important is that this book's tests will let you quickly find problems and faults early in your build. We do not guarantee passing our tests ensures your satellite will survive every disaster launch might bring, but we can say with certainty that these tests will let you quickly find and fix early stage problems that would otherwise render your satellite dead on arrival.

Choosing and De-rating Parts

If you use off-the-shelf electronics parts instead of expensive, hard-to-find space-rated gear, will your satellite work? The process of *derating* will let you do this. Engineer Amanda Shields contributes this explanation: electrical components for spacecraft have to be derated. Basically, that means that you take the electrical component and you look at the data sheet for that piece and you have to say, "Well, according to the datasheet it can have a maximum input power of XX, but NASA says that it has to be derated to 80% of that, so we can actually only have an input power of YY."

You have to do this for DoD type equipment because it is more critical for the project to succeed that a $5 mass produced alarm clock (that the owner can just send back to the company for a replacement). Traditionally they use this as the guideline for parts derating for spacecraft: *http://snebulos.mit.edu/projects/reference/International-Space-Station/SSP30312RH.pdf*. There are other options out there, Goddard has a set of requirements that are listed here: *https://nepp.nasa.gov/index.cfm/12821*

There is also a list of Qualified Manufactures and Qualified Products (*http://nepp.nasa.gov/npsl/npsl_toc.htm*). If you use an item off this list, that means it's been approved for use and doesn't have to be derated. There aren't that many listed here and it's usually cheaper to buy commercial parts and just go through the derating process.

(Amanda is working on the Capitol College VelcroSat orbital debris removal testbed picosatellite project.)

Sample Test Schedule

A sample test schedule would include:

I: Survival test and primary outgassing
> Place each board alone into chamber, run thermal vacuum cycle, remove boards and retest in isolation. This ensures that the conditions of space did not permanently destroy or distort any part of the board. It also allows for primary outgassing—the initial vaporization of any solvents, chemicals, or other contaminants on your boards. Since materials outgas in space, you want to let each board outgas on its own, so its outgassing components (if any) do not contaminate another board. See a sample schedule in "NASA Bakeout" (page 30).

II: Full system test

Re-integrate your components, ensure they work on the lab bench, make sure your battery is charged, then run a full system thermal vacuum test. This complete test lets you use your existing comm system on the satellite to monitor your spacecraft's functionality as it experiences the vacuum and thermal cycles. This is a pass/fail test. If your spacecraft passes this test, you are in great shape and ready to move on.

If your spacecraft fails this test, much like the extended analogy of a dead PC, you now have a lot of work to do isolating components and retesting. The general procedure if you fail this full system test is, first, test it fully assembled but out of vacuum.

If it doesn't work outside of the chamber, that means vacuum or thermal conditions broke something. Go back to your build-and-integrate steps, find the bad component, fix, and repeat. If your new build works, you're set! If you keep breaking the same component in thermal vacuum testing, that means a design flaw and you have a design that isn't space worthy, and thus you must redesign that component before proceeding.

NASA Bakeout

NASA has a formal *thermal-vacuum bakeout* process you must perform before integrating your payload into a NASA rocket, listed in the PPOD/CubeSat Requirements Document we cited earlier in this chapter. Their justification is "thermal-vacuum bakeouts are critical in assemblies of space flight hardware to ensure the lowest levels of outgassing." A short form of it starts with bringing your vacuum chamber (with payload inside, of course) to at least 5×10^{-4} Torr. Record the pressure level and temperature, and keep recording this every 10 minutes. Raise the temperature from room temperature (25 C) to 70 C. Let it bake at 70 C for one hour, bring it back to room temperature, wait an hour, then do a second (final) bakeout up to 70 C for an hour, then finally let it come back to room temperature. "This will eliminate most of the outgassing that will occur at this temperature extreme."

Solving Problems

Did it fail testing after you took it out of thermal vacuum? If so, there's a part that's failing due to space conditions. That is a serious problem. There are three approaches to solving this: brute force, stepwise integration, or better testing.

For brute force, you simply test your flight spare instead and hope it works. If it does, your original had some subtle manufacturing or assembly defect, so toss it aside (with red stickers indicating *DOA* so you don't accidentally use it!) and move on.

Stepwise integration is an approach you can immediately do instead of brute force, or the first thing to try if brute force failed (if the flight spare had the same thermal vacuum failure to operate). In this, you separate then slowly re-integrate each item in a series of thermal vacuum steps. A suggested sequence uses a truncated version of the original integration tests. I recommend:

1. Power + comm system to test that it can send signals (functionality retained in thermal vacuum)
2. Power + comm + CPU to test full functionality is maintained in thermal vacuum
3. Power + comm + CPU + sensors
4. All components, mounted in the skeleton: full build test

A better testing approach requires you incorporate test gear into your thermal vacuum chamber. The intent here is that you can test each piece in isolation while it is within the thermal vacuum chamber. Typically, this means creating an extra port on your chamber so you can run your lab bench test leads (whatever you used to test it earlier) into the chamber. Then, you go through the full build-and-integrate test schedule, only this time in thermal vacuum, in order to precisely determine the point of failure. Armed with this information, you can then assess whether it is a manufacturing defect or a design flaw.

Test Sequence

From this, we get our set of required tests:

1. Vacuum testing I: Primary outgassing
2. Vacuum testing II: Surviving in an airless environment
3. Vacuum testing III: Thermal vacuum testing
4. Vibration and shake testing: Simulating a rocket launch
5. G-forces: Simulating the thrust of a rocket launch
6. Drop test: Simulating a drastic quick shift in position
7. Electromagnetic interference testing: Shielding and noise response in your circuits

What Exactly Are We Testing?

If you're new to electronics, you may wonder what exactly we are testing, in a physical sense. There are three main things to check after any of these vibration, gravity, stress, bounce, and destruction tests. These are: gross physical damage, continuity, and actual operations. First, give the tested part a visual inspection to check for gross physical damage. Have any parts fallen off or become loosened? In particular, look for loose connections and loose solder joints.

If everything is tight, then you want to do a quick test with a continuity probe or voltameter to ensure that the circuits are indeed working. Working means that parts that should be connected still are, and that parts that shouldn't touch are indeed not touching (and not causing a short circuit). You can specifically test for voltage drops and voltage level at key test points in your circuit. Key test points are something your schematic diagram indicates should have a specific value (a zero reference value, or a specific measured voltage). As part of this check, ensure that all grounds are indeed solidly attached to your primary ground.

Finally, you should power up the circuit and check that it functions as it should. Test it with appropriate input to make sure it is producing the desired output. If it's tight, connected, and working, it passed the test.

Formal Risk Analysis

A "risk" is the term for something that could go wrong. A formal look at risk analysis (part of systems engineering) deals with concepts such as risk categories, bundling risks, availability analysis, and making requests for changes (RFCs). The first step is to categorize a given risk as technical, schedule, or cost risk.

Technical
  Product may not be satisfactory

Schedule
  An essential milestone may not be met

Cost
  Cost budget may be exceeded

Risks are ranged by their likelihood of occurrence (low/medium/high) versus the consequences if risk is realized (low/medium/high). This gives you the overall severity (small, medium, large), often plotted as a 5x5 matrix (Figure 4-1).

Availability analysis looks at mean time between failures (MTBF) and mean time to restore (MTTR) where availability A = MTBF/(MTBF+MTTR). For example, given a time between failures MTBF=400 hrs and a time to restore MTTR=2 hrs, we get an availability A=0.995, or during 1 year the system is unavailable 16.7 times for a total of 43.8 hrs.

		1 VERY	2 LO	3 MODERAT	4 HIGH	5 VERY
P R O B A B I L I T Y	5 VERY	L	M	H	H	H
	4 HIGH	L	M	M	H	H
	3 MODERAE	L	L	M	M	H
	2 LO	L	L	M	M	M
	1 VERY	L	L	L	L	M

Figure 4-1. *A standard risk matrix, image from the NASA Procedural Requirements.*

Going deeper, formal Failure Modes, Effects, and Criticality Analysis (FMECA) is a process. In NASA's MIKL-STD-1629A, "Procedures for performing a Failure Mode and Effects Analysis":

1. Define conditions and constraints for each component
2. Define what constitutes failure
3. Assign probabilities and severities

Once you have a failure occur—an anomaly—you need to resolve it. The fundamental rule in anomaly resolution is that an anomaly, no matter how complex, has one and only one cause. The corollary: multiple failures don't occur unless they cascade from a single root cause. From a testing point of view, this means that, if a component is not working or has a fault, you need to keep drilling down until you find a single root cause.

To give an example, if you plug in a TV and it doesn't turn on, the problem could be the power, the on button, the TV's display panel, or a variety of other causes—but it's not all of them. Your job in testing and anomaly resolution is to find the one thing that went bad and caused the failure.

Fun fact: a survey of 13 spacecraft over 2 years indicates anomalies occur about every 7 months of spacecraft operations, with 1 lost during launch. For picosatellites with a lifetime in weeks, this figure isn't useful. If your satellite is going to stay up for years, however, then you need to accept from the start that there will be anomalies.

The bulk of anomalies for satellites occur during launch and early orbit. This is the unproven time, the portion where a new satellite is experiencing the rigors of space and being powered up and operated for the first time in orbit. A short list of possible disasters to expect during launch and early orbit includes (from Squibb, Boden & Larson, *Cost-Effective Space Mission Operations*, 2006, pg 314):

- Increased volume (3x) of real-time events
- Volume of commanding (amount of instructions being received and run by the spacecraft computers)
- Amount of data processing for engineering telemetry
- Number of people involved in operations
- Number of activities that must be scheduled, checked, and conducted
- Shortened cycle times (doing more in less time; less time to move from activity to activity)
- Increased risk
- Tremendous forces on the rocket during its ascent
- Many *firsts* associated with each activity and subsystem
- Rapid sequential activities
- Single-point failures in the system design
- Demands on the ops team's reaction (human factor)
- Launch and deployment

That's a lot that can go wrong. Let's start building gear to help launch-proof and orbit-harden our satellite on the ground, while it's still in our hands.

5/Thermal Vacuum Chamber

Vacuum is a nasty environment. We have no pressure, we have outgassing, and (least we not forget) we have the simple removal of air. Materials *outgas*, which is to say, liquids and volatile solids boil off. This outgassing can then coat itself onto nearby parts. Ever buy something wrapped in plastic, and the item smells like plastic for the entire day after you unwrapped it? That's outgassing. Now imagine plasticky vapor depositing itself on, oh, your eyes, and bonding there. Because in space, the outgassing stuff just may hit parts of the satellite as it evaporates away.

That is bad. It can coat detectors or solar cells. If any outgassing materials happen to be conductive (unlikely but possible), you can mess up circuits. Plus, it's not tidy, letting stuff swirl away like that. However, in a vacuum chamber, you can have the bulk of the outgassing occur safely on the ground, sparing your components once in orbit.

Vacuum testing also ensures your soldered links are stable, and aren't going to break due to air pockets, air conducting where there should be solder, and similar mishaps. That's part of the *no pressure* and *removal of air* issues. Maybe there's a part or two that is fragile and can't take the drop to zero pressure. Maybe there are air bubbles in a component due to poor manufacturing, which will cause it to break when it hits vacuum, as shown in Figure 5-1. (As the external pressure lowers, the balloon expands due to internal pressure.)

This is why you want to test your satellite in conditions as near to where it will be to make sure it can handle space. It's pretty straightforward, actually. Imagine if you were building an underwater detector—you'd want to send it underwater. You don't have to enumerate everything that might go wrong, you just immerse and see what happens.

So it is with vacuum. Instead of imagining every possible mishap, just test the sucker. And one usually does thermal testing along with vacuum testing. Heat and cool your satellite while it's in a vacuum to mimic the space conditions you expect. Hope it survives. Repeat if necessary.

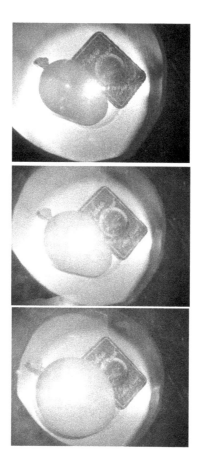

Figure 5-1. *A balloon in our vacuum chamber as we go from normal pressure to soft vacuum.*

If you're curious what happens to humans in space, here's the scoop from Imagine the Universe (*http://imagine.gsfc.nasa.gov/docs/ask_astro/answers/970603.html*). Summary: a human in vacuum doesn't explode, but does black out in about 15 seconds. Turns out our skin is like a good quality balloon; it holds us together quite well.

Pressurized components, microfine wires, and highly finicky components are a separate issue. If you are flying an item that is pressurized and must keep that pressure even in space, without leakage, then you have to do extended vacuum testing. If you have a component that is in vacuum during launch and must remain evacuated until it gets to orbit, you have a situation requiring additional shock and vibration testing.

What is it like in orbit once we get there? In many ways, being in orbit (while a challenge) is, at least, less difficult than surviving the trip to orbit. Historically, more picosatellites have been lost during launch than during in-orbit operations. However, vacuum is where the satellite will spend most of its life, so it needs to survive in an airless hot/cold environment.

Vacuum is hard to survive, as I keep emphasizing. Building a vacuum chamber can be harder. Conceptually, it is simple: an airtight chamber attached to a vacuum pump, that is strong enough not to shatter inward due to pressure differences. In practice, as with any plumbing or pressure work, you hit problems:

1. Any weak point will pop, ruining your vacuum.
2. If there is no weak point, the whole thing might up and burst.

The core orbit tests are vacuum and thermal conditions (Figure 5-2).

Pressures and vacuum notation

The only thing harder than getting a good vacuum is learning all the schemas for describing it. Pressure is often measured in Pascals (Pa), Torrs, or atmosphere. A Torr is, by definition, 1 mm of Mercury displaced under normal gravity and pressure (written as 1 mm Hg). Sometimes *inches Hg* are used instead of *mm Hg* (1 inch = 25.4 mm). A Torr is 1/760th of an atmosphere and also equal to 134 Pa. Conversely, one atmosphere is 760 Torr or 101.325 kPa (kiloPascals) or 29.92 in Hg.

Since vacuum one Earth is a negative pressure, vacuum pumps such as the COTS auto one chosen here use a scale of both mm Hg and Torr (0-30 inches Hg equating to 0-760 Torr). I used Torr as a measurement unit because that is the unit my gauge uses for calibration. With atmosphere at 760 Torr, a soft vacuum is anything below that down to perhaps 25 Torr, a good hard or high vacuum (desirable) is 1×10^{-3} Torr, and space itself is 1×10^{-6} Torr. If you have a gauge using inches, 1 atm = 29.92 in Hg.

Figure 5-2. *GSFC's large thermal vacuum chamber, big enough to fit a car. Image courtesy of NASA.*

Building and Using a Thermal Vacuum Chamber

Pumping down to -20 Torr (total pressure 740 Torr) technically is considered a *dessicant*, while pulling more than -20mm Torr of vacuum pressure are proper vacuum chambers. In absolute measures, a *soft* vacuum (or low vacuum/coarse vacuum/rough vacuum) is an absolute pressure going from just below 760 Torr atmosphere to perhaps an absolute level of 25 Torr and below, while the hard vacuum expected in LEO is a more characteristic 0.001 Torr ($1x10^{-3}$ Torr) and below, ideally getting down to 10^{-6} Torr. Achieving a stable hard vacuum is very hard, often expensive, and immensely finicky. For about $50, a half hour's work, and no serious measuring, you can have a chamber capable of fitting a CubeSat and achieving soft vacuum (est. 250 Torr).

Fortunately, the resin/modeling community has done good work in homemade low vacuum chambers. For home use, we distinguish between high vacuum chambers and low vacuum chambers. We're aiming for the latter.

Low vacuum is easier to reach and achieves most of our core goals:

- Outgassing
- Surviving vac/low pressure
- Thermal/radiative heat testing

As it happens, I ran into another "human vacuum" issue when buying the pump for my DIY chamber. For preliminary testing, I picked up a hand vacuum pump, the type often used to check automobile break lines. It's not the hardest vacuum, but it's enough to do the primary outgassing and basic sanity checks I'll need early on.

The vacuum hand pump presented an unexpected challenge, particularly when hunting on eBay. It turns out there is an, err, type of vacuum hand pump that is used for, umm... enhancing the size of the male sexual organ. I did not want (nor need) that.

Besides, those pumps look to be flimsy plastic, unable to hold a hard vacuum. And I want to test the satellite at levels that would make a typical sex organ simply explode (and not in a good way). As a matter of fact, even for automotive-grade parts, the plastic pumps are to be avoided; I went with metal.

High-grade vacuum pumps

On pro-grade oil pumps for vacuums. The good part is they can reach a higher (harder) vacuum. The downsides are they have a higher cost and, if your chamber fails, they have a backspray of oil that renders the entire chamber and all its contents oil-soaked and unusable without significant cleaning.

In the end, I bought a simple brass *brake bleeder* hand vacuum pump at Harbor Freight. On sale for $20, bargain! And best of all, I can safely certify that my satellite is 100% sex gadget free. Still keeping our PG rating, woo!

I hunted up DIY vacuum chamber construction ideas from others, and the best set of informational links was at DVForum. They use vacuum to make bubble free wax. Okay.

Among the other materials I need is a primary retort, and material I can use to make a lid. The lid will be plastic—thick acrylic. As it happens, I learned how to work acrylic when making custom pickguards for the electric guitars I like to build.

So a technique from building electric guitars comes in handy when making a musical satellite. That shows that DIY hobby work is just like being in orbit. What goes around comes around.

Any good plan needs a fallback. If my DIY isn't sufficient to pass InterOrbital spec, I can always hit up a nearby university to see if I can borrow their chamber—it's a lot easier to borrow something if you've already tested the payload in your chamber so it's not likely to explode in theirs.

The $100 Thermal Vacuum Chamber

The components (and one tool required) are:

1. A cheap 8qt (or bigger) stainless steel pressure cooker
2. A 1 foot x 1 foot sheet of Lucite or similar transparent plastic
3. A 1 foot x 1 foot 1/8" thick 60A smooth rubber sheet
4. A hand-pump *auto/brake line bleeder* vacuum pump
5. A brass hose fitting
6. Drill with spade bit able to make hole for hose fitting (typically 3/4")

7. Use of a tap-and-die to make screw threads for 3/4" hose hole

The main engineering required is that you'll throw out the top of the pressure cooker to make your own, using the rubber sheet as a soft washer to make a seal and the plastic to provide rigidity.

You put the rubber sheet on top of the pot, then the rigid plastic on top of that. Drill a hole in the plastic and a corresponding hole in the rubber for where the pump will attach. Cut out an *observation window* in the rubber—but make sure to leave a larger border (at least 1") around where the rubber meets the top of the pot. Drill, thread, and install the hose fitting into the plastic. Hook up the pump and you're ready!

By the numbers:

1. Throw away the top of the pressure cooker and just keep the pot.
2. Lay the rubber sheet on top of the pot.
3. Lay the Lucite on top of the rubber. Decide where you want to drill for the hose fitting, and where you'll want to have a see-through observation window.
4. Cut a hole in the rubber sheet in the places corresponding to the hose fitting.
5. Cut the rubber sheet to make the observation window, again keeping at least a 1" border around where the rubber meets the top of the pot to ensure a good seal.
6. Drill the hole in the Lucite for the hose fitting, thread the hole, then screw the hose fitting in (including a rubber washer, which is usually included in the vacuum pump kit).
7. Using the hoses provided in the vacuum pump (and following that kit's set-up instructions), attach the pump to your new hose fitting.
8. Make sure everything is clean and dry.
9. Lay the rubber on top of the pot, the Lucite on top of the rubber, and begin pumping (Figure 5-3)!

The rubber sheet will have an initial outgassing, so do your early tests and cycle the chamber (evacuate then let come back to normal pressure) several times to allow this outgassing, before you even think of putting your gear in it.

On Pressure Cookers

Can't you just use the pressure cooker *as is* for vacuum? Some people on the Web have reported using a pressure cooker as-is to maintain vacuum, and that is very tempting since it includes a hose fitting in the steel top. However, the existing sealing mechanism is designed to be tight only when there is high pressure inside, and when you instead do the negative pressure of a vacuum, it's leaky. Also, pressure cookers have a latch/valve combo on the handle that seals only if there is positive pressure. Although you can drill out the latch and manually hot-glue seal the valve, you still find that the included gasket requires positive pressure to seal, and is not useful for a vacuum. There may be some brands of pressure cooker that are vacuum-ready, but if so, I haven't found them.

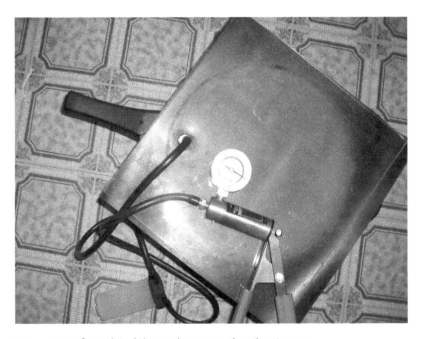

Figure 5-3. *Completed thermal vacuum chamber in use.*

Parts List

There are the parts I used, plus comments on variants you can easily get. The total came to roughly $75,

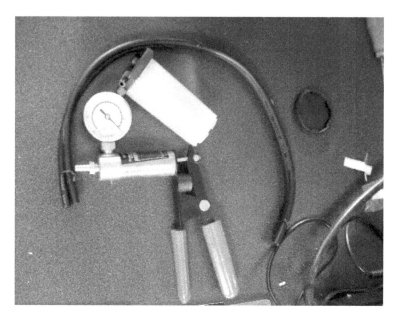

Figure 5-4. *COTS Brake bleeder hand vacuum kit.*

Brake Bleeder and Vacuum Pump Kit
$24.99 at Harbor Freight or online or just about any auto shop. The pump is surprisingly easy to find from an auto parts or hardware store. This type is typically used to bleed automotive break lines and can get down to -40 mm Hg, which is higher capability than our setup requires. Includes the vacuum pump, intermediate chamber, fittings, washers, and has a vacuum release to let you gently bring it back to normal pressure (Figure 5-4).

8qt Pressure cooker
Mirro 8qt Pressure Cooker, Aluminum, from *http://amazon.com's* Warehouse Deals, $24.90 (Figure 5-5).

A thick rigid plastic 12"x12" sheet
Polycarbonate or high impact acrylic, 3/16" minimum thickness (1/2" preferred, and thicker is fine) (Figure 5-6). Acrylic is said to be more prone to shattering, resulting in shards of acrylic shrapnel if it does shatter. That said, I went with the high-impact Lucite-Tuf acrylic because I had some from Home Depot from an earlier project (Lucite-Tuf being more shatter-resistant than regular Lucite). Specifically, I used Lucite-Tuf High Impact Acrylic Sheet, 3/16" (0.177", 4.5mm) thickness (but thicker is better). Note that Lucite, Plexiglas, and Perspec are all acrylics, while Lexan is one brand of polycarbonate. Polycarbonates

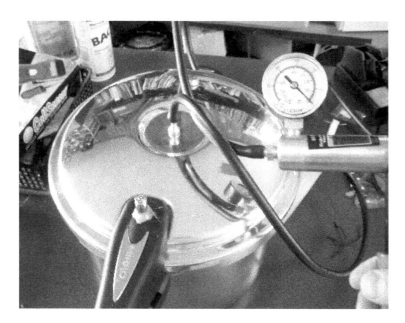

Figure 5-5. *COTS pressure cooker as main chamber*

typically are about twice the cost of acrylic. A 12"x12" 1/2" runs about $18 for polycarbonate or $11 for lucite (both *http:// usacustomplastics,com*, plus $2.50 shipping). I recommend checking your local hardware store and/or a glass shop.

One Medium-Strength Neoprene Rubber Sheet, No Backing
60A Durometer, Smooth, Standard, ASTM D2000 BC, Black, 12" Width, 1/8" Thick, 12" Length", from *http://amazon.com*, $9.93. I suspect I could have gone with a slightly thinner sheet but there was no strong price difference (under $1 difference). Likewise, I settled on a hardness of 60A as it is a common rubber hardness. I would not go harder than 70A, as you need some degree of compression to form the vacuum seal. There's a handy table at *http://mykin.com/rubber-hardness-chart* that compares hardness types, for those wishing to experiment with more optimized seals. They range from *rubber band* (20A) to *shopping cart wheel* (100A). Let us know what you find out!

Brass 1/2" Barb adapter
$2.00 (Lowes/local hardware shop), a standard fitting that has a 1/2" thread and a 1/4" *nipple* to attach the hose.

For a design involving more milling and work, but using the same concept, a hacker named AllanJ has this design:

http://dibblah.pwp.blueyonder.co.uk/VacuumChamber/

Figure 5-6. *Sheet of rigid plastic, being drilled for port*

And there is supplementary comments on Hack A Day include general tips:

http://hackaday.com/2011/12/23/a-vacuum-chamber-from-a-pressure-cooker/

Chamber Performance

We are using a hand pump, so this can be a little tiring. It took just 1 minute and 100 pumps to reach -130 Torr (-5 in Hg); 5 minutes of pumping to reach -380 Torr (-15 in Hg); another 5 to reach -500 Torr (-20 in Hg). Fifteen minutes of pumping reached -560 Torr (-23 in Hg), at which point I believe the chamber reached its limits. Past that, I could maintain but not increase the vacuum level. Engineers making a chamber with higher fidelity can certainly improve on this design, however, we repeat that a $50 chamber that reaches soft vacuum and allows for outgassing and desiccation is an ideal component for any picosatellite builder.

For holding the vacuum, when set to an initial vacuum of -500 Torr (-20 in Hg), it dropped to -380 Torr (-15 in Hg) after 1 hour, and by 7 hours was at -130 Torr (-5 in Hg). If you want to do a lengthy outgassing or longer testing, you simply need to pump more. For those curious on the decay rate, using approximate conversion factors (it's not that accurate a scale, after all):

Table 5-1. *Vacuum pumping times*

Minutes elapsed	Torr	inches Hg
0	-500 Torr	20 in Hg
5	-460 Torr	18 in Hg
10	-430 Torr	>17 in Hg
20	-430 Torr	<17 in Hg
30	-400 Torr	>16 in Hg
50	-400 Torr	<16 in Hg
60	-380 Torr	15 in Hg
210	-260 Torr	10 in Hg
300	-200 Torr	7.5 in Hg
420	-130 Torr	>5 in Hg

Thermal/Vacuum Testing

Briefly, there are three ways that heat works in Earth. You can gain heat or lose heat (cool off) through conduction, convection, and radiation (radiative heating and radiative cooling).

Conduction
> The heat flows from hot region to cold for a given solid, liquid, or gas. This includes flowing through anything touching your payload. If your payload has a different temperature than its surrounding air (or liquid), it will equalize. The material doesn't change or move, the heat just diffuses from hot to cold.

Convection
> The heat moves via currents in the material. Think of heat causing the material to flow away, carrying more heat than conduction.

Radiative
> This is heating up by absorbing photons (light) or cooling by radiating photons (light). For example, laying in the sun on a beach during the day, you absorb the Sun's visible photons and heat up. Once it gets dark, you radiate out that acquired heat as infrared light and cool down. Color

affects the rate of radiative heating and cooling. In general, darker colors both absorb photons more readily, and radiate that heat back with equal strength if there is no light. Light or silvered objects absorb less, but also radiate less.

There is no conduction or convection in space, because there is only vacuum around you—no material to conduct heat, no material that could have currents to carry away heat. Only radiative heating and cooling occurs in space. In orbit, your problem is, further, that you alternate being in hot sun and cold dark. You tend to either heat rapidly, or cool rapidly, without having any buffer between the two states.

For example, NASA notes at "Staying Cool on the ISS" (*http:// science.nasa.gov/science-news/science-at-nasa/2001/ast21mar_1/*) that "without thermal controls, the temperature of the orbiting Space Station's Sun-facing side would soar to 250 degrees F (121 C), while thermometers on the dark side would plunge to minus 250 degrees F (-157 C)."

A lighter color moderates the heating/cooling rate, so you get less hot in the sunlight and you hold the heat a little better so you get less cold in the dark. This is why the shuttle is painted white; it gives it better thermal control. Adding insulation also slows down the rate that you heat and cool, and this in turn reduces the extremes in temperature you will experience. If you can slow down the rate of heating, for example, then you don't get as hot before the next dark cooling period begins.

For thermal/vacuum testing (literally testing the thermal properties while in a vacuum), I had two internal setups. One was to place the piece being tested directly on the base of the chamber. As an offset, I used a metal *mini anvil* as a platform. Given the size and scope of our chamber, I do not think there is a strong difference in either outgassing or vacuum testing with either approach. However, they will behave differently for thermal testing.

In thermal testing, there are (in the absence of air) only two heating and cooling mechanisms available: conductive and radiative. True LEO vacuum has only radiative (nothing there to conduct), but for our chamber, you have the contact point between the piece and the chamber itself to allow you more direct thermal conduction.

I recommend a direct approach for thermal tests, using external heat and cold sources conducting through the chamber walls directly to the test piece in order to test the thermal response of the pieces under temperature extremes. Choose the temperature extremes you expect from our earlier discussion (such as -30 C to 120 C), you can easily get to 0 C using ice or -78C using dry ice, and a high of 100 degrees simply with a boiling water bath, or higher using a hot plate or gas burner.

For this, you simply heat (using a burner) or cool (using ice or dry ice) the outside of the chamber near where the work is placed. Assuming you place the piece directly on the floor of the chamber or on a platform attached to the floor, that means just heating or cooling the base. You do not want to provide strong heating or cooling to the top of the chamber, as that may make the rubber seal brittle or leaky and damage your chamber.

You can heat or cool the bottom of the chamber, and the portion of the piece inside that is touching the chamber or platform will then heat (or cool). By varying this, you run the piece through its temperature extremes.

Heating Rates Due to Sunlight and Dark

For an object in Earth orbit, the sun puts out 1361 Watts per square meter (called the *solar constant*), assuming you're facing it straight on. Assuming a 10cm x 10cm Cubesat profile, that means it gets about 14 Watts of direct heating on its surface. Alternately, you can call the solar constant 0.13 Watts/cm^2. We want to mimic this to see how our payload's temperature will vary in space over time.

For a pure radiative setup, you can offset the object from the base using some sort of platform. You must minimize the amount of *junk* inside the chamber, so avoid plastics or foam as much as possible. Otherwise, your platform will outgas and contaminate the work! If you want a platform, the metal mini-anvil plus a thin foam or rubber coating to thermally isolate would suffice.

To provide purely radiative heating, shine a hand-held incandescent 100-Watt light bulb over the display window to provide heating of the top side of the material. We are estimating that a 100 Watt bulb in a reflector housing (so the bulk of the light is directed out) is spread over the approximately 1/3 meter diameter chamber yielding about 100 Watts over 706 cm^2, or 0.14 Watts/cm^2—pretty near our required solar constant of 0.13 Watts/cm^2. A more formal test should also use ultraviolet (UV) light, which requires special bulbs (*blacklight* bulbs from novelty stores are a source of low power UV light, in a pinch).

The bottom side and any parts not in direct contact with the light will experience much less radiative heating—and that is only from reflections from the chamber's interior. Any part that is in darkness will have no radiative heating, and will only feel heat due to slow conduction through the piece itself.

To ensure only radiative cooling, cover the observation window and chill the outside of the chamber (surrounding it with ice, for example). The item will emit infrared light, cooling in the process. The reason for chilling the chamber is to ensure that any emitted IR that impinges on the chamber walls will be carried to the outside, and that the chamber itself does not emit IR to head the object. Radiative cooling is a very slow process.

For these tests, you want to monitor how quickly the temperature rises or falls when it is in sunlight or in darkness. This will give you an idea of how the payload temperature will vary. If you run a pair of 90-minute orbit simulations, you can gain a very good idea of the likely temperature range it will experience as well.

To run this, just place the object in the chamber and evacuate to soft vacuum. Keep the chamber dark, surround it with ice, and let it cool to a low equilibrium of a bit above 0 C. This is our estimate of the initial cold temperature your spacecraft is likely to experience. As we noted earlier, your spacecraft will swing between temperature extremes over the course of a 90-minute orbit, and part of this test design is to try and discover what those limits will be for your satellite's exact shape and materials.

To run the thermal test, shine the radiative light source on the payload for 45 minutes (half an orbit), then go dark for 45 minutes, then light it for 45 minutes, then go dark for 45 minutes. The temperature change over time for that second pair of light/dark should give you a decent estimate of how quickly your satellite heats and cools, and what limits are likely.

These limits derived from simulated orbit day/night cycles are the temperature operating range that your satellite must be able to function within.

6/Launch Tests

Historically, a number of CubeSat and conventional satellites fail to reach orbit or fail upon arrival into orbit. The stresses of a rocket launch are the number one killer of satellites. Preparing for the initial launch and early operations phase of your mission is therefore the most critical in determining whether your mission succeeds.

However, rocket testing is harder than in-orbit testing. You can cook up a thermal vacuum chamber to test vacuum and temperature swings expected during the mission lifetime. It is harder to simulate a rocket launch. We will walk through some testing to prepare. The testing you need to do is primarily shake, rock, and roll.

Shake tests are vibration and acoustic testing to ensure your satellite can handle the rattle and hum of the rocket launch. Rock-and-roll testing is the centrifuge G-force testing to simulate your satellite being crushed by G-forces as the rocket barrels up in flight. A drop test neatly wraps up the set, simulating small, sharp movements during prep, launch, or in deployment. All three are ways your satellite can be battered about any of its axes.

Key to note is that, since picosatellites are often a secondary payload, you may not know which way will be *up* for your satellite launch. You cannot build assuming a specific orientation. Therefore, you need to test how your satellite survives G-forces along all directions.

At a recent nanosatellite conference, I asked whether NASA or other places could assist with testing for survivability on a rocket. The answer was mixed. Basically, there are places that have facilities for test, that you can either beg time or pay for time to use them. This is almost always on a *who you know* basis, however. You need to contact or network with people at a well-equipped university that does picosatellite work, in order to use their shake-rock-and-roll testbed.

Vibration Testing

Michael Collins of Apollo 11 wrote about his earlier Gemini rocket flights, "The first stage of the Titan II vibrated longitudinally, so that someone riding on it would be bounced up and down as if on a pogo stick. The vibration was at a relatively high frequency, about 11 cycles per second, with an amplitude of plus or minus 5 Gs in the worst case." (source: *http://www.vibrationdata.com/rocket.htm*) The Constellation program had a primary typical frequency of 10-15 Hz (cycles per second).

Figure 6-1. *The van-sized Astro-E X-ray satellite is put on a shake table, with 100 accelerometers attached to measure stress and strain as it is put through the shake test.*

A rocket vibrates as it launches, due to engine throttling and (during the early parts) wind resistance. Each stage of the rocket will induce changing vibrations, as well as the vibration from stage shutdowns and stage separations. The range of vibration depends on the rocket.

The NASA Preferred Reliability Practices report (Practice No. PD-ED-1259) gives the typical range as being from 20 Hz to 2,500 Hz, with carry-over frequencies up to 10,000 Hz. This wide a range is perhaps beyond home testing.

Vibration testing can be done on a rocker table, shaker table, or other apparatus to mimic these cases. Many professional testbeds can do frequencies up to 10,000 Hz with a 2-inch throw (moving the object back-and-forth 2 inches over 10,000 times each section!) (Figure 6-1). We're going to aim for a shorter throw and focus on the lower frequencies in our home testing.

Vibration Spectra

Vibrations have both *frequency* and *amplitude*. *Frequency* is how many times per second there's a shake, and *amplitude* is the strength of each shake. To give an example, if I tap you three times each second, that's 3 Hz in frequency (3 times per second, since 1 Hertz = 1 Hz = once/second) with a low amplitude. If I punch you three times each second, it's the same frequency but with a stronger amplitude, and you will therefore take more damage.

Each rocket has its own vibration spectra, which is a profile of the range of frequencies and amplitudes that your payload on that rocket will suffer. The NASA spec mentioned in Chapter 4 provide one set of numbers. Alternately, this handy reference from the CalPoly "Test Pod's User's Guide" to the NASA GEVS *worst case* profile: (*http://www-personal.umich.edu/~mjregan/ MCubed/Pages/Documents/TestPodUser%27sGuide.pdf*) gives you a set of limits. If you can survive the GEVS worst case, you can survive any rocket out there.

Table 6-1. *GEVS Worst Case Vibration Spectra*

Frequency (Hz)	ASD Level (G^2/Hz) Qualification	Acceptance
20	0.026	0.013
20-50	+6 dB/oct	+6 dB/oct
50-800	0.16	0.08
800-2000	-6 dB/oct	-6 dB/oct
2000	0.026	0.013
Overall	14.1 Grms	10.0 Grms

The GEVS table from the *Test Pod User's Guide*, California Polytechnic State University CubeSat Program, Jonathan Brown, 2006, document, public document

The CalPoly reference also includes the schematics for a *test PPOD*, or test rig that simulates the brief shock of deploying from your rocket to space. For those seeking specific hardware tests, I recommend you consult that document and also request the vibration spectra from your rocket launch company. Again, our focus in this book is DIY testing to improve the robustness of your satellite and increase your chances of surviving launch, successfully deploying, and surviving orbit for several months. Ideally, even if you have access to a university, corporate or government test facility, you will want to do the tests in this book to ensure your satellite has greater odds of passing their more thorough (and more expensive) testing regime. Either way, you win.

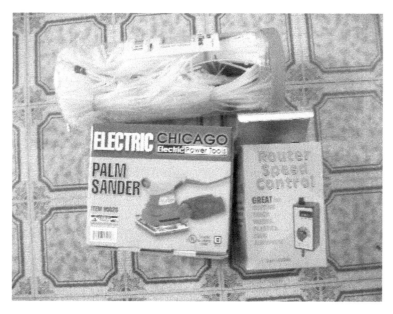

Figure 6-2. *All you need for a homebuilt short-throw vibration table: an oscillator sander, a workbench, cable ties, and a router speed control.*

Shaker Design

For our inexpensive homebuilt shaker table design, I modified an electric orbital sander to hold the spacecraft, then added a variable speed control (rheostat). You can also get variable speed orbital sanders. Look for a random-orbit or oscillating sander, typically available for $25-$60. The frequency for these can range up to 15,000-25,000 rotations-per-minute (rpm, e.g., with a DeWalt random orbital sander), which yields approximately 250 Hertz or higher (or 250 cycles per second) of vibratory oscillation. To this you'll add a router speed control, nylon, cable ties, and something to bolt it to (Figure 6-2).

By design, these tend to be fixed-speed devices, so you'll also need to purchase a *router speed control*, which is a 120V AC potentiometer designed to make any power tool variable speed (Figure 6-3). Being able to dial this down yields a range of 10 Hz to 250 Hz, giving you an inexpensive test of the primary resonance vibrations that you have to survive during launch.

The sander needs a solid base. I used the same $25 vise workbench used for the homemade centrifuge. Vise the sander in, then add a shield around at least three sides of the device in case parts go flying off during testing (Figure 6-4). Seriously, this setup will put out very potent shakes, and you need to minimize potential collateral damage if a part breaks loose.

Figure 6-3. *An off-the-shelf router control will let you dial down the fixed-speed sander to cover the range needed. (image courtesy of router manufacturer)*

Figure 6-4. *A sturdy workbench holds the sander, and a cardboard shield protects bystanders from flying debris.*

Finally, you need to fasten the payload to the vibration pad. Use two sets of nylon cable ties. The sander bed uses a lever that is usually there to tighten down on the sandpaper. Since we are not using sandpaper, instead you open the lever. Then, fasten the payload with one set of cable ties so it's slightly tight. Practice closing the level—tightening the cable ties—until the ties are at maximum tightness with the lever closed.

Figure 6-5. *Use double twist ties to ensure the work doesn't fly across the room and smash into a window.*

Now, attach the second set of cable ties to the fixed payload and hand-tighten. Make sure that the two sets of cable ties are applying their force to different sections of the payload, so that you have two points of stability where they are fixed to the sander.

At this point, you are ready to test. As with any of our home-built power tool setups, make sure you stay below the apparatus, so that any parts that break up fly away safely overhead.

Run the vibration cycle slowly from low- to high-frequency, then dial it back down. I recommend you run it for at least 1 minute at each of a half dozen frequencies, doubling each frequency shift: 10 Hz, 20 Hz, 40 Hz, 80 Hz, 160 Hz, 250 Hz. Assume a linear response, that is, given a frequency of 10 Hz at the lowest setting and 250 Hz at the highest, you can estimate where each frequncy interval is (approximately). During the transition between frequencies, do this as slowly as patience allows.

Acoustic Testing

A NASA report definition states that "acoustic noise results from the propagation of sound pressure waves through air or other media. During the launch of a rocket, such noise is generated by the release of high velocity engine exhaust gases, by the resonant motion of internal engine components, and by the aerodynamic flow field associated with high speed vehicle movement through the atmosphere."

Figure 6-6. *The acoustic test chamber at NASA/GSFC, big enough to fit a truck (or a satellite). Figure courtesy of NASA.*

Acoustic testing is similar to vibration testing (as it is sound-induced vibration) combined with pressure testing. In this case, the acoustics can cause pressure increases on the payload. Acoustic waves up to 160dB are used in professional tests. Space Travel (*http://bit.ly/OsoiuX*) notes a testing device that spans a frequency range of 20-5,000 Hz at pressures (amplitudes) up to 160dB. If you wish to do industrial-grade testing, however, that is beyond the scope of this document.

I cannot recommend doing 160dB acoustic testing in a home lab. At NASA's GSFC, their acoustic testing uses speakers taller than a person, do not allow anyone in the room, and use tremendously thick doors to ensure sound leakage out of the room doesn't deafen bystanders. The American Hearing Research Foundation (*http://american-hearing.org/disorders/noise-induced-hearing-loss/*) lists 15 minutes at 115dB as auto horn/loud rock concert level, causing damage in just 15 minutes of exposure. At 140 dB (jet engine, gun shot), damage occurs even if exposure in brief.

If you have a sine-wave generator hooked up to a normal stereo amplifier and wish to sweep, briefly, across a 20-5,000 Hz frequency range at low amplitudes (below 100dB, aka *chainsaw level*), that should suffice for picosatellite work. Remember that the acoustic vibrations are already covered in your vibration testing, and the pressure differentials set by acoustics are less for a compact picosatellite than it would be for the extended body of a larger satellite, ergo, neglecting acoustic testing should not greatly increase your risk.

Drop Testing

This is the easiest test. It is designed to mimic any of the several short, sharp movements your payload will encounter, including:

- An accidental drop during transport or integration
- The small *kachunk* drop that occurs when your payload is placed into the payload compartment
- The small *kachunk* as the payload is attached to the rocket
- Any less-than-gentle placement of the rocket onto the rocket launch stand
- Any short impulses that occur as your payload is ejected from the rocket during deployment

You can see why it's very useful to have a payload that can survive a brief short shock. I once saw a NASA/JAXA satellite do a drop test. It was anticlimactic. The entire payload was in a stand, then **poof** the whole thing dropped a very short distance. You can mimic this with the famous one-foot drop test used by computer hardware people. Take your payload, lift it a foot above your desk, then drop it. If it can't survive that (from any facing), redesign.

A more formal drop test can be done to ensure that any momentum transfer from weight surrounding your picosatellite doesn't add to the impact. Add a flat five pounds of weight on top of the payload, and drop from one foot (again). This adds *crushing* to the previous drop's "impact shock".

Most picosatellites tend to be either a titanium CubeSat shell or strong plastic tubing braced against boards using steel screws (like TubeSats). I've had a friend stand on a CubeSat shell and it hasn't buckled. Use the drop test to make sure that everything is extremely firmly fastened into place and tightly held together.

A common formal spec is to have a safety factor of 1.25 or more. Given our on-the-fly calculation style and use of estimation, though, that's 1.25 of an already vague number. For now, we go with *test it hard*.

Note this is all completely different from a *drop test* used to test parachute landings or payload recovery. Consider this the launch-drop or accidental-drop testing, not a recovery drop test.

Static Loading

Larger satellites are tested for stress and strain—forces pulling unevenly upon them. For picosatellites, being small and compact is an advantage. A picosatellite performs as a nearly perfect rigid compact body, and builders can often neglect issues of whether there is a different strain on one end versus the other. Picosatellites with extended components—expanding solar panels or robot arms—will be deploying such items only after launch, in zero-G conditions, so testing as a rocket test component is not necessary here.

We now handle the last portion of our rocket test, building a centrifuge to see how our picosatellite handles the crushing thrust forces of launch.

7/G-Force Testing

Rockets are powerful stuff, and satellites and astronauts experience tremendous G-forces pushing down on them during launch. Your satellite will experience up to 10G (10 times Earth's gravity) during launch. It is important to test this. For picosatellite work, it is necessary that your design be able to withstand these forces. To test this, the easiest way is to build a centrifuge. It is also easy to test. We'll offer two systems—the lab-oriented motor centrifuge and a more casual rope centrifuge. Both can reach up to 3G force.

Think of the spinning bucket gimmick. If you tie a bucket to a rope and fill it with water, you can make the bucket swing in a loop-the-loop over your head and not spill, as long as it is spinning fast enough. You need enough spin to counteract the 1G of the Earth's pull, so you need a spinning centrifuge of at least >1G.

Most homebuilt centrifuges are used by mad scientists doing medical or chemical studies that require they separate a fluid by density. For satellite building, we can use a far simpler design. Yes, you can actually just spin your satellite in a bucket at the end of a rope—or build a 3G spinning rig using an ordinary electric drill.

G-Force Theory

For both designs, you are going to spin your satellite on the end of a long rod or rope in order to create centrifugal force to mimic high G forces. This is how exciting amusement part rides can spin you and force you back into your seat (or why you get thrown against the door during a very fast automobile turn).

The governing equation is that `force = mass * acceleration = mass * gravity`. Since we're trying to create pseudogravity by the acceleration due to spinning, we use the centrifugal term where `acceleration = velocity squared / radial distance out`, or our `angular velocity (rotations per second) squared * radius`.

Put simply, if you double your spinning radius for a given spin rate, you double the G force you're creating. If you double your rate of spin, your G forces goes up by a factor of four. So we can look at four setups:

Electric Drill
1. Attaching a half-meter throw arm to an electric drill
2. Attaching a longer meter throw arm to an electric drill

Swinging a Rope
1. Swinging a payload at the end of a 1 meter rope by hand
2. Swinging a payload at the end of a 2 meter rope by hand

We have to look at the specifications for electric drills both in terms of how fast they turn (rpm) and how much torque or twisting power they can provide. We'll need both—we are generating false gravity by the spin rate, but we need the torque to actually move our spinning test rig at all. An ordinary 1/4" electric drill typically spins at 1,600-2,800 rotations per minute (rpm), or 24-48 Hz (rotations per second). A 3/8" electric drill provides more torque in return for a penalty of a slightly slower spin rate of 300-1,600 rpm. Finally, a heavier-duty 1/2" drill drops to 500 rpm or under 10 Hz, in favor of higher torque.

For our rig, we chose a 3/8" drill simply because we found a 1/4" drill did not have enough torque to move our centrifuge rig. Torque is crucial, in particular because the drill has to move not just the payload, but everything attached to it. We are applying a heavy load by attaching a centrifuge swing arm, dropping the speed by an estimated 50%. We also have to factor in the weight of the centrifuge arm. For each kg of our desired payload, we will end up having several kg of centrifuge arm to support this.

Alternately, we can simply use a *bucket on a rope* method and fasten our payload to the end of a sturdy rope, then spin that rope by hand to achieve our requisite force. How many rotations per second must we achieve? Let's look:

- 1 Gravity is 9.8 m/s^2, which we'll round to 10 m/s^2 for convenience.
- Our pull in Gs = w^2 * radius (meters), where w = radians per second.
- One rotation is 2pi radians, so Gs = (2pi)^2 * rps^2 * radius (meters), or approximately 36 * rps^2 * radius.
- Assuming a 1 meter throw arm, then our pull in Gs is (36 * 1m/10 m/s^2) * rps^2 = 3.6 * rps^2.

To achieve (for example) 10 Gs, we need 6 rps (36 = rps^2)! If we reduce our throw arm to 0.5 m, we need just over 8 rps (72 = rps^2). If we increase our throw arm to 2 m, we need just over 4 rps (18 = rps^2).

Experimentation shows that a typical 3/8" drill can manage 3G using an 1 meter throw arm, while a sufficiently slow build-up can lead to spinning a 2-meter rope at 3G.

Danger! By definition you are spinning a heavy object at very high speed. If your apparatus breaks, you basically have shrapnel flying around your lab. Do not do any 10G centrifuge work in the presence of people, animals, windows, or breakables. If your 1-meter long throw arm breaks during a full speed test, that object is going to fly off with a velocity equal to dropping that object on your head from a second story window. It's basically a 1kg weight traveling at the speed of a shotput. So be careful and remove living creatures and breakables.

A typical 3/8" electric drill can spin from 300 to 1600 rpm (rotations per minute), providing a torque (or spinning power) of 500-100 Newton meters, i.e., at 300 rpm you get 500 Nm, at 1600 rpm you get 100 Nm. Often drills say they can get only half their usual rpm under a heavy load, which we'll factor in later. A drill rpm * a conversion factor of (60sec per min) yields the rps, or put simply, rps = rpm * 60. For proper geometric calculation, you convert rps to the odd unit of *radians per second* by multiplying by 2pi. Torque, or twisting force, is given as Torque = Force * length = (for this case) F * arm_radius.

This actually makes a handy lab for a physics class. For the simple case of a 1-meter rotational arm, we calculated the necessary rotations per second earlier. Torque is more key—do you have enough twisting power to spin it? Given `torque = G * radius` (where "G" equals whatever G force we are trying to achieve) and we're using a lever arm of radius=1m, you need to be able to pull 10 N*m of torque—again within the limits of most electric drills. That is just for the payload, though, and neglecting the mass (moment of inertia) of the arm itself.

Roughly, for each kilogram of apparatus that must spin, you can multiply the torque necessary by that factor. So if we need 10 N*m of torque per desired G-force for a 1kg payload, adding on 4kg for a sample apparatus means we'll need `4*10` N*m of torque (4x the amount). So to pull 3G, we'll need `3 * 4*10` N*m of torque, or 120 N*m. Remember now our factor of 50% efficiency if the drill must operate at high torque, and suddenly we're at 300 N*m of torque needed. Assuming a 1-meter arm and thus needing 6 rps (aka 360 rpm), that means we're just about at the limits of our drill.

A writeup at the Aquatic Pathobiology Laboratory suggests plotting a nonogram of RPMs versus G-forces needed as a visual comparison of G-force (maximum relative centrifugal force) to RPMs (rotations per minute). Overkill for this level work, but it's worth introducing the term for those who wish to study further.

Hand-Powered G-force Rig

As mentioned, you can simply place your payload in a bag at the end of a 2-meter rope, then go outside and start swinging the rope around. I was able to reach G-forces (see sidebar for using a smartphone to measure it) of 2.5G with simple manual spinning. While this may lack a certain coolness, it's a very quick and easy way to do some initial structural tests.

Accelerometers

You will want to calibrate your system to see how many Gs you can pull. Almost all smart phones now include an accelerometer, as do a number of gaming devices. An accelerometer is a device that measures how much you move (or accelerate) your device in any direction. It's a fancy term for a type of internal motion sensor. There are several iPhone and Android apps for measuring G-forces, including listings under *accelerometer*, *elevator*, and *Porsche*. I recommend you run the rig with a smartphone running in an accelerometer mode (Figure 7-8) to see how high a G-force you are achieving.

Many amateur high altitude balloon projects have flown an iPhone or Android phone as their balloon payload. Some picosatellite builders are contemplating the next step of flying an iPhone as their primary satellite command and communication system. With the growing number of sensors being built into smart phones, your smart phone could also service as a science device. For now, though, we'll just use it for assistance in testing.

The usual warnings apply: do this in an open area, clear the area of anything that an escaping payload might damage, and clear all bystanders out of the area. One plus is, since you're the motor at the center of this setup, you're in the safest place possible. You will want to adjust the length of the rig to find a personal sweet spot. I found a 20-foot rope had too much drag to spin fast enough, while a 5-foot rope was easily spun but needed much higher RPMs to get a decent G-force (Figure 7-1). For my rig, a middle length of 12 feet worked best, but of course it's both easy and more practical for you to find one that works for your own level of personal strength.

The Drill-Powered G-force Rig

The components you need are very simple: a spade bit to use as the axle, wood blocks to secure the axle, a long piece of wood as a throw arm, glue to fasten it together, a workbench to brace it, a drill to power it (Figure 7-2).

Figure 7-1. *An extremely simple hand-spun G-force rig that takes about 1 minute to build and can reach 2.5 G without trouble.*

Figure 7-2. *Everything needed to build your own 3-G centrifuge.*

First you need to build the swing arm that fastens to the drill. We use a 1" wood boring spade bit as our axle. We'll take advantage of the flat head of this bit by encasing it in wood blocks (Figure 7-3). We can then attach our long wood arm to this axle and have a stable propellor-like centrifuge swing arm.

To build the housing for securing the spade bit, first lay the three pieces of wood as if they were a solid block and examine how the bit will fit. In our case,

Figure 7-3. *A snapshot of the axle during assembly, showing the layers.*

we drilled a hole for the base block, a slot for the middle long arm, and a hole +half slot for the top block. This way, the three blocks will solidly trap and encase the spade bit and create a very solid connection. Glueing the three blocks will create a solid connector (Figure 7-4).

Once built, you now have a propellor-like swing arm that will fit into any standard drill. We need to put it onto our 3/8" drill and find a way to secure this rig so it's stable.

You need a very solid base. I used an inexpensive ($25) vise worktable to serve as the centrifuge base. The drill clamps vertically into the table, allowing for a free horizontal spin path for the centrifuge (Figure 7-5). Your rig doesn't have to be perfectly aligned or balanced, but aim to get close to that. Setting the axle by hand sufficed, and I found (for 1kg payloads) I didn't need to add a counterweight to the other end. Warning—a homebuilt centrifuge can seriously hurt you.

You can't run this setup by holding the drill in your hand, obviously. I picked up a $20 workbench to fasten this to. Ideally, you should run this outside, using tent stakes and rope to ensure the base is secure. Place the drill inside the vise-top, then tighten it down. Most drills have cheap plastic casings so it can be hard to find a stable part to vise into. I found near the drill chuck was more solid than the body. Once it's vised in place, realize it is now fixed along one axis (left to right) but can still tilt up or down slightly, as the

Figure 7-4. *The axle, glued.*

Figure 7-5. *The completed rig, ready to spin.*

vise is not directly preventing that direction. The resulting wiggle along that direction was extremely strong and therefore requires more support. Clamping in crossbars was sufficient to create a stable hold along both directions (Figure 7-6).

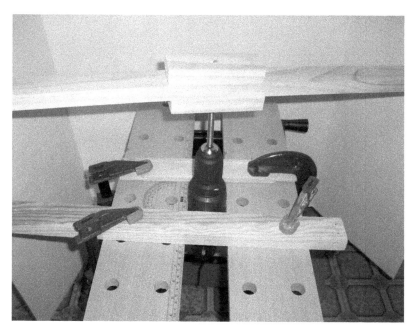

Figure 7-6. *Firmly brace your centrifuge, using the vice-like table for one axis and either clamping or (for a more permanent rig) nailing down crossbars to support the other axis.*

To fasten the payload, I drilled holes at the beam's end, then used twist ties to secure the payload onto the bar. Pictured is a test setup using a bare TubeSat shell (Figure 7-7).

Drill: you need a variable speed drill, the sort that goes faster the more you push the trigger. You need to start slow then let it gradually come up to speed. A $25 3/8 inch drill from Harbor Freight sufficed for me. I did try it with a fixed-speed drill; the jerk where the drill tries to go from *zero* to full speed nearly flipped the rig over. So go variable and go slow. The safest place to be during a run is underneath it, which puts you nicely in place to run the test.

1. Set up rig securely, and fasten in payload.
2. Give it a test spin to make sure it's roughly balanced.
3. Crouch underneath and slowly bring it up to speed.
4. Run it a minute or so, then let it slow down.
5. Remove and examine.

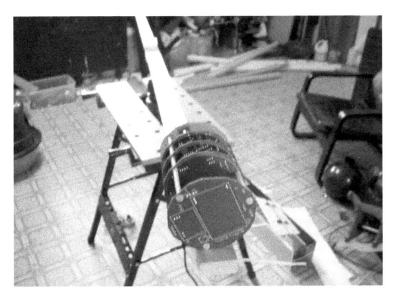

Figure 7-7. *A sample payload on our homebuilt 3-G centrifuge.*

With our heavy test setup, we were able to achieve 3G with the standard 3/8" electric drill. At that level, though, the drill engine actually smoked. It still ran, but it was clearly at its operational limit. A better design would minimize the weight of the axle mounting blocks and the centrifuge throw arm, as they weighed significantly more than the payload and thus the bulk of the drill's torque was not being used by the payload. However, even a 3G rig that costs under $50 and can be built in 20 minutes will be of great use in doing your initial rocket gravity testing.

Avoiding Harm to People and Things

Two of the pieces of test gear (G-force centrifuge and Shake table) use ordinary power tools (electric drill and oscillating sander, respectively) but can do significant damage to a workshop or person if there is a malfunction and the user doesn't take precautions. For the G-force rig—basically a giant horizontal propeller with the payload attached the end—if it breaks loose from the vise mounting or the actual rig or payload detaches, it will fly out at great force and can break a window (or person, if someone is careless enough to stand anywhere other than where I recommend, which is underneath the rig in the operator's station). For the shake tester, we're vibrating the payload at high oscillations and a part could conceivably break free (or the entire payload could become unattached) and go shooting across the room (I recommend safety goggles, placing it in a cardboard enclosure, and as always staying in the operating station underneath the rig.)

By analogy, it's about the same danger as using a chainsaw to cut down trees —not a problem if you follow procedures but not quite as routine as drilling a hole.

Figure 7-8. *Centrifuge spinning a securely taped down iPhone to measure acceleration.*

8/Good Test Procedures

The answer to where you should test your picosatellite is *everywhere*. You need to test it in the lab while it's being built, then after it is being built. Then test it remotely from across the lab, hands-off, to make sure the basic communications channels work. You can then fly it on a helium low-altitude balloon to make sure it works *floating free* and that you can communicate with it and operate it. Try a high altitude balloon if you can, to see how it behaves in rarified air. Booking a sub-orbital or ballistic launch can give it a good rocket test and vacuum test, though it's worth pointing out that will make it experience extra jarring.

Your satellite will by nature be modular in design. It will have the basic bus or platform (one component), then the instruments or experiment. A given bus design can be reused for different experiment packages, so having a robust bus that survives one flight is a great way to start a series of picosatellite missions, each building on the last. Finally, picosatellites come in 1U (single cube), 2U (two cubes), and 3U (3 cube) form factors, and you can consider testing them in segments.

Flight Spares

You must always build at least two of every spacecraft board. I cannot emphasize this enough. During testing, you will have a primary build set that you are using, and your flight spare. Both should be tested. By definition, your flight model will be the board that best passes testing, and the *spare* is the second best of the boards that survived testing. (Any board that fails testing is simply DOA—dead on arrival—and should be thrown out or reworked.) Further, if a board fails on your primary at any stage of the test, you can swap in the working flight spare and be able to proceed without having to restart at square one.

This can be illustrated by example. A designer (call her Gallant) builds two of each board. It's not much work to build two at a time, since all the equipment and gear is already set up. Once done, she can put away the current gear and then set up her lab bench for the next step, so everything is moving in a timely, well controlled fashion. Each of her boards test well in isolation, and also when tested solo in a vacuum chamber. While doing a final test of the satellite in the vacuum chamber, however, it doesn't work, and examination reveals a capacitor split open. Gallant can simply take the tested flight spare, swap it into the integrated build, and retest in the vacuum chamber. It passes, so she's ready to move on.

Meanwhile, a second designer (call him Goofus) builds only one of each board. Again they test well in isolation on the bench and in the vacuum chamber. But when his integrated build fails, he now has more work. He has to rebuild the defective board, and test it on the bench, and retest it according to the integration test, then retest it solo in vacuum. Each time, he has to remember how he did the original, reset his work environment, and redo all those steps. Only then can he swap in the new component and get back to the integrated testing. Poor Goofus has blown his build schedule by not having a flight spare.

Often, you'll make several of a key component—a CCD, a collimator, an amplifier. Then you'll test them all, and put the highest performing one onto the final payload. So the spares aren't *bad*, per se, but they are not as optimal as the ones you use (obviously enough).

Should you do more than two? Heck yes. If you can afford it, build three or four of a given component. For expensive items, such as solar cells or Radiometrix transmitters or (depending on your budget) Arduino or BasicX-24 chips, you may be limited to two. However, you can still get 3-4 of a given PCB printed and place the stock components, neglecting only the expensive parts, so that you can do a quick swap-and-go refit if a primary board fails due to a solder joint or a non-expensive component.

The pluses of using flight spares is that the hardware research is already done, as is most of the manufacturing. So the cost is much lower. The downside is, a flight spare isn't a ready-to-fly backup. Instead, it's one or more partial assemblies of the equipment. Typically, it hasn't been tested—or it's the parts that were tested and failed, so they got shuffled out of the final payload. So flight spares can be a source of parts, but slightly lesser parts.

For those that are future-looking, having working, tested spares around also means your second and third satellite are very evolved after your first launch is a success. No one said you only have to fly once!

Test Scheduling and Isolated Testing

A good test schedule is folded into your build and integration timeline. You are not going to go straight from building a satellite on a lab bench to launching it into orbit. Part of testing is testing each component in isolation before you hook it up to the other parts. Then you test the entire build while it's on the ground. From that, you do distance tests to ensure both that the satellite components work and that, equally important, you are able to remotely send and receive sufficient health and safety information that you know whether it works.

Next, an optional but recommended step is to do a short balloon launch, using an easy-to-get weather balloon and store-bought helium. This test is like a warmup run, a full shake-out that has a little bit of shaking and moving, a little bit of distance command and control, and the tiniest bit of pressure work. A balloon that goes up only to the height of a typical model rocket launch, say 1100 feet (335 meters), is only going 1/1000th of the height you need to get to orbit, and will experience only a 4% pressure drop from normal atmospheric pressure. However, it experiences all those changes—shake, distance, and pressure—at the same time. One way to think of the balloon test is that, while it will not ensure success, it can let you quickly diagnose any major problems cheaply and easily, well before you launch into space and have no recovery or recourse available.

For such low altitude ballooning, you can use a tether line to ensure you can recover your balloon. For more extreme altitude balloon tests, issues such as tracking, range safety (i.e., avoiding hitting houses, powerlines, and aircraft), and recovery of your balloon are a topic for a separate book. One dubious advantage of rockets is that *recovery* of our picosatellites is easy—there isn't any recovery, they burn up on reentry.

Even before we get to testing your entire satellite, though, we must first test each piece as it is being built. There are two categories of problems you will run into while testing: component failure and integration issues. Component failure means that a given component or board simply does not work. It could have a bad part, an error in assembly, or an incorrect design. Whatever, it just doesn't work. Integration failure, in contrast, is when two working parts do not work when put together.

Integration errors are the more frustrating part, because they can be harder to isolate and diagnose. Therefore, you want to test each individual part in isolation before you put the parts together.

Imagine that you bought a new PC and you hook up the computer to a monitor, keyboard, and mouse—and the darn thing shows a blank screen. As a system, you have very little information other than "it doesn't work". The first thing most technical people do is try to isolate the problem. Is the monitor on and working? Does the keyboard and mouse work on a different PC? Is the computer powering on? Does it work with a different monitor? In essence, you are testing each component to determine, isolate, and localize the failure. With that, you know which one part to repair or replace.

Since you are building everything one piece at a time, you can save yourself many "it doesn't work when put together" problems by testing each piece before you marry it to another. *In isolation* means you test that piece on its own, using whatever power or computer or network connections it may require to operate. For example, testing the communications radio in isolation means you hook up your satellite radio board to a spare battery and a temporary antenna and see if it sends and receives signals. You can use any signal generator or a PC or an Arduino board or a BasicX-25 board or any similar tool to handle the inputs and outputs. Only after you have tested the comm unit in isolation, to ensure it works, do you marry it to your actual flight CPU for further testing. This way, any problems in one unit are easily diagnosed.

Suggested Build-and-Test Schedule for a Typical Picosatellite

- Power bus (using a rechargable battery)
- Solar cells (isolated test)
- Solar cells + power bus to ensure the solar cells can charge the battery (then put aside the solar cells for now, as they are fragile)
- Communications (radio) as a stand-alone unit, using a temporary antenna
- Communications using the power bus
- Communications using the flight antenna (then put the antenna away for now, as it is a speciality part)
- Controller/CPU in isolation
- Controller/CPU using power bus
- Controller/CPU and communications together using power bus
- Sensors and instruments and other payloads in isolation
- Sensors and instruments and other payloads using power bus
- Sensors and instruments and other payloads using power bus and CPU
- Sensors and instruments and other payloads using power bus and CPU and communications
- Sensors and instruments and other payloads using power bus and CPU and communications while housed in your satellite skeleton

At the end of all of this, you have a satellite!

Defects

Defects can exist because:

1. You ran into a bad commercial-off-the-shelf (COTS) part. Swap it and move on.

2. A COTS part failed due to bad tolerances or an incorrect choice of a COTS part (tolerances not sufficient for operations in space). This is partially

a design issue; a part rated at +/-10% might have worse performance in space or at temperature extremes, or the given part is not rated to the temperature range expected in space so you need to choose an alternative part with better tolerances.

3. Failure in assembling your board (such as a bad soldering joint). Rebuild and retest.

4. Design failure (as built, the board fails in space due to a fundamental issue). Tricky. Design failures are the worst because you have to go all the way back to your pre-build stage and do the entire process over again.

Frame Testing

Mount the components into the satellite skeleton and ensure they still work. Yes, we are suggesting you not put the boards into your pretty CubeSat or TubeSat frame until they all work. This saves wear and tear on the mounting hardware. If all your components work when put together on your lab bench, odds are good you only have to put them into your satellite skeleton once. On the other hand, if you try to keep installing the boards into your skeleton before testing them solo, you may end up screwing and unscrewing the pieces so often your screws and screw holes and mounting hardware begins to get worn. Worn hardware will fail launch testing. So remember —test in isolation, test together, then test as a satellite.

Alternately, if you have more than one skeleton (as with both the CubeSat and TubeSat examples here), by all means feel free to use one skeleton as a test rig and keep a second, pristine skeleton for your final assembly. Even in this case, I recommend you do as little mechnical assembly as possible until all the parts are tested.

Rocket science, in the end, is more about risks than easy answers.

Clean Rooms

A *clean room* refers to a room that is clean, usually kept that way by having filtered air pumped in and maintaining positive pressure so no other outside air can enter. Clean rooms can fit a satellite and the team needed to build it (Figure 8-1), but for picosatellite work, we have to aim smaller.

My clean storage device of choice is a rolling trunk (Figure 8-2), and I'll admit it's not the most exciting science prop. But there are four important details that make this an important detail in DIY satellite building.

Figure 8-1. *A car-sized satellite (ASCA) in a NASA clean room.*

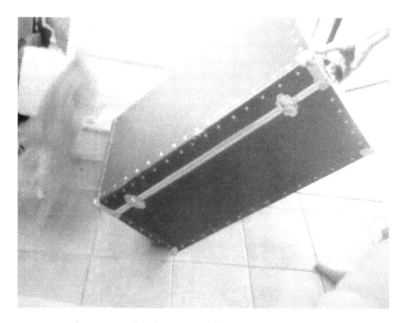

Figure 8-2. *Any setup that lets you isolate your equipment from contaminants is a positive step.*

Pragmatics
> I need a place to store the satellite and components when I'm not working on it.

Conveyance
> I need something to lug the kit around when I take it over to my friend's workshop, when we eventually get to final assembly.

Promotion
> I need to be able to lug the satellite (or flight spare) to classrooms when I talk to K12 students about cool rocket science. The wheels are a crucial help here.

Cats
> Notice the blur of the fast-moving cat to the left of the rolling trunk? Yes, we have a cat infestation problem in our house.

Back when I worked at Naval Research Lab, we had cats—strays that roamed the garages and underbrush around the facility. But we never had cats indoors, and certainly did not have cats in our assembly rooms and labs.

But being stuck with a home lab means I must somehow ensure the cats don't mess up the satellite. I've declared our cat-free guest room as my lab, so I'm not worried about cats knocking stuff down or—worse—rewriting then recompiling the sensors code. No, my worry is fur.

I want my picosatellite to be the first musical satellite, not the first fur-covered non-functioning cat satellite. This means I have to set up a clean room. A clean room in rocket parlance is exactly what it sounds like—a very clean room. Typically, this means taking a room that is spic-and-span, and creating positive pressure. You make sure that clean air is being pumped into the room, and rely on the fact that the air has to leave as your way of keeping dust out.

A common setup is to run powerful intake fans that are well filtered, so they pump and filter lots of outside air in. You surround it with plastic curtains, so the air can escape, and rely on the mild wind (for want of a better term) to ensure no outside dust enters.

DIY Clean Room/Box

For bonus points, you can make the scientists wear bunny suits—that's the actual name of the white cleanroom outfits. There are varying degrees of clean room standards. A large clean room can be built using the "DIY Clean Room" instructions at the fine *http://instructables.com* site. For our work, we aren't doing fine optics or detectors with a high contamination risk, so we need only a moderate system.

Over at Make, they have something more my speed (*http:// blog.makezine.com/2006/10/how_to_make_clean_box_low/)—a* homemade clean, err...clean box. It's really more my speed. It's just what it looks like—a plastic tub covered in transparent plastic, with little rubber gloves inside that you access from holes on the outside. But even that may be overkill for our picosatellite, because, again, we have no optics, sensitive electronics, or other easily contaminated items.

In previous NASA missions, I've seen—on rare occasion—flight hardware sitting in the middle of dusty labs. Most hardware is fine that way, because most of it is a big hunk of metal. The only time for clean rooms is final assembly, when you put in the sensitive optics. You do vacuum tests later, and outgassing usually removes most contaminants. Vibration testing, likewise. Wrap it up and you're done.

Cleanliness is a progression, not an absolute from square one. So first I'll just be in a clean, cat-free room. Later, I'll use the trunk to transport to the metal shop. Then back to the lab. I'll be building a vacuum chamber and vibration rig for testing. After that, I will transfer it to a sealed bag for final shipment. The result will be a clean, but not ultra-clean, functional satellite.

Someday, you picosatellite builders may get to work on a major project, such as shown in this panorama of the JWST being assembled in the world's largest clean room (Figure 8-3). Note the engineers in bunny suits at the far corner.

Figure 8-3. *NASA's largest clean room, at GSFC, for the JWST. Image courtesy of NASA.*

The Money Build

There exists an industry of pre-made satellite boards. Much as you can assemble Lego Mindstorm or other kits of parts to build a robot as purely an integration exercise, so it is with Cubesats. You can buy, mix, and match. This leads to the existence of the "money build", where the *do it yourself* aspect is lessened and the kit-bashing operation of buying pre-made components and entire satellite subsystems is maximized. In short, you're spending money to avoid building things yourself.

The question this raises is why you are building and launching a satellite. There are many reasons (covered in the previous book, *DIY Satellite Platforms*), but from a construction point of view, we can break this into two reasons. You either want to learn to build satellites or you don't care about the satellite bus, just the payload.

Using an existing set of boards will teach you how satellites work. Likewise, making satellite boards from any of the various levels of *from scratch* is fine, whether that means designing your own circuits, fabricating using existing schematics, or buying pre-made kits. It's all relevant experience and you get to choose how deep you want to delve.

I'll argue a money build is an excellent exercise. There are three reasons not to just buy pre-fabricated parts. The first is that you want detailed experience in electrical engineering or satellite electronics concepts, those skills you only get by building from scratch. Bear in mind that everyone has a different level of *from scratch*. For some people, integrating existing boards is a sufficient design challenge. Other people would be satisfied with nothing less than creating your own schematics (no pre-existing circuit board layouts!) and then using only a rocket you engineered yourself. Anything in between is up to you—DIY means you get to choose your own path.

The second reason you may not want to use pre-fab components is that you want to build a satellite to do or test something for which a pre-fabricated board does not exist. In that case, you have the option of using pre-fab for standard components (radio, say) and focusing your design skills on the novel parts.

The third and most common option not to use pre-fab components is cost. At $2K-$8K per board, you can easily spend $50K on your satellite build. Adding to this, if you want to do a series of satellites (each advancing on the previous design), each satellite will have that same base cost. Designing, fabricating, or assembling in-house, by contrast, gives you a lower cost-per-satellite simply because it's not significantly more work to create four assembly-line style boards than it is to solder up your first, so you gain economy of scale.

My recommendation is not to be motivated by ego ("must do it all solo") and instead approach it from a systems engineering perspective. Using pre-fab pieces reduces risk because they are tested and proven solutions, but increases material cost. Pre-fab parts will likely reduce your labor costs—reduce the time it'll take you to build. Factor in how much you value your time and what aspects of the project are most enjoyable or educational for you, they balance the key trade-off of cost versus risk and cost versus time.

In my thoughts, you're as authentic a DIY satellite builder whether you slapped together CubeSatShop components to fly a Sputnik-type signaler, or smelted your own copper to build your own circuits. The DIY is "yourself" and your goals, no one else's.

When to Commit

I got into an argument with a NASA engineer over the use of clean rooms for picosatellites. His claim was that you cannot build a satellite outside of a clean room. My stance is that it is perfectly fine to build and test your picosatellite in a typical air-conditioned sealed non-industrial lab. By *non-industrial*, I mean a lab set aside for the satellite work and not containing any machinery that puts out smoke, oil, or metal shavings. No building your satellite in a metal shop—build components there, clean them, then bring them to your (reasonably) clean lab bench.

Although I did not build hardware for the NASA missions under which I helped do science work, I did get to see some of the flight hardware and flight spares. For a small mission in the early stages, due to pragmatics you will often assemble them in a lab. You are trading off the cost and speed advantages of using an ordinary lab, against the necessary fidelity of the instrument. Under this, your need for clean room assembly is partially driven by the quality and complexity of your design.

If you are creating James Webb Space Telescope-level mirrors at a cost of many millions of dollars, you are justified in also building the world's largest clean room to assemble these. These are extremely large yet extremely high quality optics. However, if you are building a 6cm x 6cm x 6cm satellite to prototype a new controller chip or set of store-bought components, your requirements can relax the *clean room* standards to using just, well, a room that is clean (as opposed to a vented positive-pressure *clean room*).

Either way, at a certain point you need to shift from your lab/bench world to the pristine environment of pre-launch. A picosatellite is supposed to be developed on a short timescale using simple tools. As one builder put, if you're spending five years and hundreds of thousands of dollars on your picosatellite, you're doing it wrong. But with a bit of boldness and a strong do-it-yourself spirit, you can be flying (Figure 8-4) your own picosatellites "the maker way".

Figure 8-4. *Beautiful deployment of the Atmospheric Neutral Density Experiment (STP-H2-ANDE4). This could be you!*

About the Author

Alex "Sandy" Antunes is an astrophysicist who turned to science writing upon realizing the desire to understand the universe doesn't mean you have to be the one to discover everything personally. There's a lot of excellent science out there, and Sandy enjoys bringing it to the world's attention. Sandy recently achieved a professorship at Capitol College's Astronautical Engineering department, which he credits to his NASA work, his solo build of the Project Calliope picosatellite, and his writing for Science 2.0 and via O'Reilly Media.

Surviving Orbit the DIY Way is the second book in the four-book DIY Satellite series. The first, *DIY Satellite Platforms*, covers the basics in designing and building your own satellite. Book 3 will be on instruments and science you can do with your own satellite, and Book 4 will provide how to communication with and operate your picosatellite once it is safely up in space.

Built with Atlas. O'Reilly Media, Inc., 2012.

Have it your way.

O'Reilly eBooks

- Lifetime access to the book when you buy through oreilly.com
- Provided in up to four DRM-free file formats, for use on the devices of your choice: PDF, .epub, Kindle-compatible .mobi, and Android .apk
- Fully searchable, with copy-and-paste and print functionality
- Alerts when files are updated with corrections and additions

oreilly.com/ebooks/

Safari Books Online

- Access the contents and quickly search over 7000 books on technology, business, and certification guides
- Learn from expert video tutorials, and explore thousands of hours of video on technology and design topics
- Download whole books or chapters in PDF format, at no extra cost, to print or read on the go
- Get early access to books as they're being written
- Interact directly with authors of upcoming books
- Save up to 35% on O'Reilly print books

See the complete Safari Library at safari.oreilly.com

O'REILLY®

Get even more for your money.

Join the O'Reilly Community, and register the O'Reilly books you own. It's free, and you'll get:

- $4.99 ebook upgrade offer
- 40% upgrade offer on O'Reilly print books
- Membership discounts on books and events
- Free lifetime updates to ebooks and videos
- Multiple ebook formats, DRM FREE
- Participation in the O'Reilly community
- Newsletters
- Account management
- 100% Satisfaction Guarantee

Signing up is easy:

1. **Go to: oreilly.com/go/register**
2. **Create an O'Reilly login.**
3. **Provide your address.**
4. **Register your books.**

Note: English-language books only

To order books online:
oreilly.com/store

For questions about products or an order:
orders@oreilly.com

To sign up to get topic-specific email announcements and/or news about upcoming books, conferences, special offers, and new technologies:
elists@oreilly.com

For technical questions about book content:
booktech@oreilly.com

To submit new book proposals to our editors:
proposals@oreilly.com

O'Reilly books are available in multiple DRM-free ebook formats. For more information:
oreilly.com/ebooks

O'REILLY®

Spreading the knowledge of innovators   oreilly.com